The Imperial College Lectures in
PETROLEUM ENGINEERING

Topics in Reservoir Management

Volume
3

The Imperial College Lectures in
PETROLEUM ENGINEERING

Topics in Reservoir Management

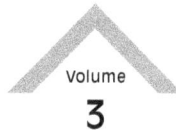

Volume
3

Deryck Bond
Kuwait Oil Company, Kuwait

Samuel Krevor
Imperial College London, UK

Ann Muggeridge
Imperial College London, UK

David Waldren
Petroleum Consulting and Training (PCT) Ltd, UK

Robert Zimmerman
Imperial College London, UK

World Scientific

EW JERSEY · LONDON · SINGAPORE · BEIJING · SHANGHAI · HONG KONG · TAIPEI · CHENNAI · TOKYO

Published by

World Scientific Publishing Europe Ltd.

57 Shelton Street, Covent Garden, London WC2H 9HE

Head office: 5 Toh Tuck Link, Singapore 596224

USA office: 27 Warren Street, Suite 401-402, Hackensack, NJ 07601

Library of Congress Cataloging-in-Publication Data

Names: Muggeridge, Ann.

Title: Topics in reservoir management / by Ann Muggeridge (Imperial College London, UK),
    Sam Krevor (Imperial College London, UK), Robert Zimmerman
    (Imperial College London, UK), Deryck Bond (Kuwait Oil Company, Kuwait),
    David Waldren (Petroleum Consulting and Training (PCT) Ltd, UK).

Description: [Hackensack] New Jersey : World Scientific, 2017. | Series: The Imperial College
    lectures in petroleum engineering ; volume 3 | Based on lectures that have been given in the
    world-renowned Imperial College Masters Course in Petroleum Engineering.

Identifiers: LCCN 2016055436 | ISBN 9781786342843 (hc : alk. paper)

Subjects: LCSH: Oil reservoir engineering. | Oil fields--Production methods. |
    Petroleum industry and trade--Management.

Classification: LCC TN870.57 .M84 2017 | DDC 622/.3382--dc23

LC record available at https://lccn.loc.gov/2016055436

British Library Cataloguing-in-Publication Data

A catalogue record for this book is available from the British Library.

Desk Editors: Dipasri Sardar/Mary Simpson/Shi Ying Koe/Jennifer Brough

Typeset by Stallion Press
Email: enquiries@stallionpress.com

Preface

This book is the third volume of a set of lecture notes based on the Master of Science course in Petroleum Engineering that is taught within the Department of Earth Science and Engineering at Imperial College London. The Petroleum Engineering MSc is a one-year course that comprises three components: (a) a set of lectures on the different topics that constitute the field of petroleum engineering, along with associated homework assignments and examinations; (b) a group field project in which the class is broken up into groups of about six students, who then use data from an actual reservoir to develop the field from the initial appraisal based on seismic and geological data, all the way through to eventual abandonment; and (c) a 14-week individual project, in which each student investigates a specific problem and writes a small "thesis" in the format of an SPE paper.

The Petroleum Engineering MSc course has been taught at Imperial College since 1976, and has trained over a thousand petroleum engineers. The course is essentially a "conversion course" that aims to take students who have an undergraduate degree in some area of engineering or physical science, but not necessarily any specific experience in petroleum engineering, and train them to the point at which they can enter the oil and gas industry as petroleum engineers. Although the incoming cohort has included students with undergraduate degrees in fields as varied as physics, mathematics, geology, and electrical engineering, the "typical" student on the course has an undergraduate degree in chemical or mechanical engineering, and little if any prior exposure to petroleum engineering. Although some students do enter the course having had some experience in the oil industry, the course is intended to be self-contained, and prior knowledge of petroleum engineering or geology is not a prerequisite for any of the lecture modules.

The complete set of lecture notes will eventually consist of about a half-dozen different volumes, covering topics such as petroleum geology, fluid flow in porous media, well test analysis, reservoir engineering, and core analysis. The present volume contains chapters on four topics that fall under the rubric of "reservoir management": rock properties, enhanced oil recovery, reservoir simulation, and history matching. Each chapter has been written by a lecturer who is either an academic based at Imperial College, or a practitioner working in the oil industry, and who has taught these and other modules within the Imperial College MSc course for many years. Although the volumes are lecture notes, and consequently do not aim to achieve encyclopaedic coverage, or to contain extensive reference lists, taken as a whole they contain the basic knowledge needed to prepare students to work as reservoir engineers in the oil and gas industry.

Robert W. Zimmerman
Imperial College London
July 2017

About the Authors

Deryck Bond holds BSc and PhD degrees in Physics from Imperial College London. He has worked in research, reservoir engineering and petroleum engineering functions for a number of major and mid-sized oil companies and as an independent consultant. He has contributed to the Petroleum Engineering MSc program at Imperial College. He is currently a Senior Consultant working on the Greater Burhan field for Kuwait Oil Company.

Samuel C. Krevor is a Senior Lecturer in the Department of Earth Science and Engineering at Imperial College London. His research group investigates the physics and chemistry of processes of multiphase flow in engineered and natural geological systems. He received his BSc, MSc, and PhD in Environmental Engineering from Columbia University.

Ann H. Muggeridge holds the Chair in Reservoir Physics and EOR in the Department of Earth Sciences and Engineering at Imperial College London. Her research interests are focussed on the numerical modelling and upscaling of reservoir flows with a particular emphasis on enhanced oil recovery processes (including miscible gas injection, low salinity waterflooding, polymer flooding, steam injection and Vapour Extraction (VAPEX)) and the impact of geological hetero-geneity on those processes. She is also interested in the mixing of reservoir fluids after reservoir filling but before production as an indicator of reservoir compartmentalization. She has a BSc (Hons) in Physics from Imperial College and a DPhil in Atmospheric Physics from the University of Oxford. She worked for 5 years at BP and 4 years at SSI (UK) Ltd as a reservoir engineering and reservoir simulation specialist before joining Imperial College. She is a member of the EAGE and the SPE, chairing the organizing committee for the EAGE IOR Symposium 2017 and 2015 and sitting on the organizing committee for the SPE Reservoir Simulation Symposium 2017. She is also an InterPore Council member and sits on the Scientific Advisory Committee for the National IOR Centre of Norway.

David Waldren is director of Petroleum Consulting and Training Limited and a Visiting Professor to the Centre for Petroleum Studies, Department of Earth Science and Engineering, Imperial College London. He has been a consulting reservoir engineer for over 25 years, previously having been with BNOC, Intercomp, and IPEC. He is recognised worldwide for his reservoir management and simulation advice. He has a PhD in particle physics from Liverpool University.

Robert W. Zimmerman obtained a BS and MS in mechanical engineering from Columbia University, and a PhD in rock mechanics from the University of California at Berkeley. He has been a lecturer at UC Berkeley, a staff scientist at the Lawrence Berkeley National Laboratory, and Head of the Division of Engineering Geology and Geophysics at the Royal Institute of Technology (KTH) in Stockholm. He is the Editor-in-Chief of the *International Journal of Rock Mechanics and Mining Sciences*, and serves on the Editorial Boards of *Transport in Porous Media* and the *International Journal of Engineering Science*. He is the author of the monograph *Compressibility of Sandstones* (Elsevier, 1991), and co-author, with JC Jaeger and NGW Cook, of *Fundamentals of Rock Mechanics* (4th edn., Wiley-Blackwell, 2007). He is currently Professor of Rock Mechanics at Imperial College, where he conducts research on rock mechanics and fractured rock hydrology, with applications to petroleum engineering, underground mining, carbon sequestration, and radioactive waste disposal.

Contents

Chapter 2. Introduction to Enhanced Recovery Processes for Conventional Oil Production   47

Samuel C. Krevor and Ann H. Muggeridge

Chapter 3. Numerical Simulation   **109**

Dave Waldren

Chapter 4.   History Matching   **209**

Deryck Bond

Chapter 1

Introduction to Rock Properties

Robert W. Zimmerman

*Department of Earth Science and Engineering,
Imperial College London, Kensington, London SW7 2AZ, UK*

1.1. Introduction

The most important fact about reservoir rocks is that, by definition, they are not completely solid, but rather are *porous* to one degree or another. The degree to which they are porous is quantified by a parameter known as the *porosity*. The fact that the rocks are porous allows them to hold fluid. If these pores are interconnected, which they are in most rocks, then the fluid is able to flow through the rock, and the rock is said to be *permeable*. The ability of a rock to allow fluid to flow through it is quantified by a parameter called the *permeability*. As the porosity controls the amount of oil or gas that the rock can hold, and the permeability controls the rate at which this oil or gas can flow to a well, these two parameters, porosity and permeability, are the most important attributes of a rock, for reservoir engineering purposes.

It would be very advantageous to petroleum engineers if the pore space of a reservoir were completely filled with hydrocarbon fluid. Unfortunately, this is never the case, and the pores always contain a mixture of hydrocarbons and water. The relative amounts of oil, gas or water are quantified in terms of parameters known as the fluid *saturations*. These saturations are in turn controlled by the surface interactions between the rock and the various fluids, which can be described and quantified in terms of properties known as *wettability* and *surface tension*.

The ability of a rock to store fluid, and the relationship between the amount of fluid stored in the rock and the pressure of the fluid, is related to the porosity, and specifically to the way that the porosity changes as the pore pressure changes. The relationship between porosity and pressure is quantified by an important mechanical property known as the *pore compressibility*.

Aside from properties such as porosity, permeability and compressibility, which are obviously of crucial importance in reservoir engineering, there are other petrophysical ("petros" ▬ rock, in Greek) properties that are important, but for less obvious reasons. One such property is the *electrical resistivity*. Although the resistivity is not directly related to the oil, its value is controlled mainly by the amount of water in the rock, and so knowledge of the electrical resistivity gives us valuable information on the relative amounts of oil and water in the rock.

In this introductory chapter on rock properties, we will define the various parameters mentioned above, present some simple models to relate these properties to the pore structure of the rock, and give some indication of how these properties are used in petroleum engineering.

1.2. Porosity and Saturation

1.2.1. *Definition of Porosity*

If one looks at a typical cylindrical core of rock, with radius R and length L, it would have an *apparent volume*, or *bulk volume*, of $V_b = \pi R^2 L$. However, on a smaller scale, such as under a microscope (Fig. 1.2.1), it would be clear that some of this volume is occupied by rock minerals, and some of it is void space.

We can now define the mineral volume of this core, V_m, as the volume that is actually occupied by minerals. Lastly, we define the pore volume, V_p, as the volume of the void space contained in this cylindrical core. These volumes are related by

$$V_b = V_m + V_p. \tag{1.2.1}$$

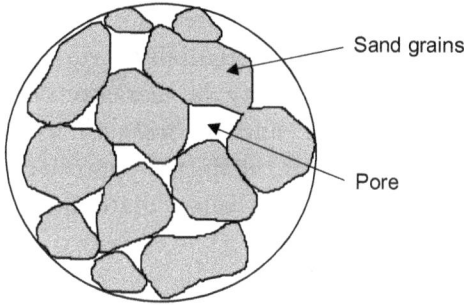

Figure 1.2.1. Schematic diagram of porous sandstone, showing grains and pore space. Typical grain sizes are tens to hundreds of microns.

The fraction of the cylinder that is occupied by pore space is known as the *porosity*, and is usually denoted by ϕ (although sometimes by n):

$$\phi = V_p/V_b. \tag{1.2.2}$$

The porosity of a reservoir rock can range from a few percent, to as high as 40%.

A distinction is often made between *primary porosity*, which is the porosity that the sandstone (say) had after it was first deposited and compacted, and *secondary porosity*, which is any porosity that is subsequently created through mineral dissolution, mineral deposition, fracturing, etc. One very important type of secondary porosity is the porosity contained in *natural fractures.* Many reservoirs, which collectively contain roughly half of all known oil reserves, are naturally fractured. These reservoirs are often filled with an interconnected system of fractures. The porosity contained in these fractures is usually in the order of 0.1–1.0%, and is much less than the primary porosity, but the fracture network typically has a much higher permeability than the unfractured rock, often by several orders of magnitude. These types of reservoirs are referred to as *dual-porosity* reservoirs. Producing oil from such reservoirs is more problematic than producing oil from unfractured reservoirs.

Another distinction that can be made is between *total porosity*, which is essentially the porosity that is defined by Eq. (1.2.2), and *effective porosity*, which measures only the pore space that is interconnected and which can potentially form a flow path for the hydrocarbons. The total porosity is therefore composed of effective/interconnected porosity, and *ineffective*/unconnected porosity. In most rocks, there is little ineffective porosity. One important exception are carbonate rocks called diatomites, in which most of the porosity is unconnected. The Belridge oilfield in central California has produced 1.5 billion barrels of oil from a diatomite reservoir that has porosities ranging from 45% to 75%, most of which is unconnected and not "effective"! But this is a special case that requires special production methods, and is not typical of the reservoirs that will be the focus of most of this course.

1.2.2. *Heterogeneity and "Representative Elementary Volume"*

The property of porosity introduces an issue, that of *heterogeneity*, which is important for all petrophysical properties. To say that a reservoir is "heterogeneous" means that its properties vary from point-to-point. In one sense all rock masses are heterogeneous, because, as we move away from a given rock at a given location, we will eventually encounter a different rock type. For example, some reservoirs contain beds of sands and shales, with thicknesses in the order of a few metres, such as in Fig. 1.2.2. These reservoirs are heterogeneous on length scales in the order of tens of metres.

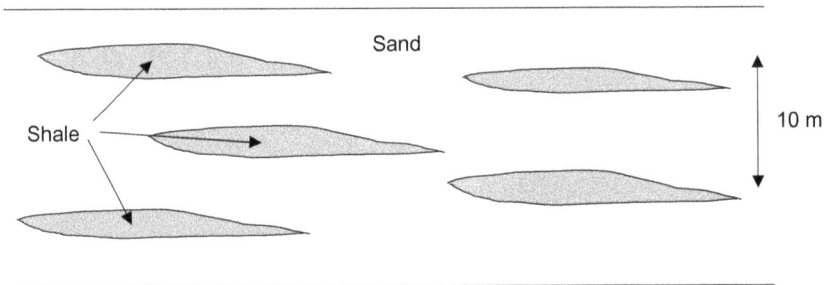

Figure 1.2.2. Sand–shale sequence in a reservoir. Such a reservoir is homogeneous on a length scale of a few centimetres, but heterogeneous on a length scale of a few metres.

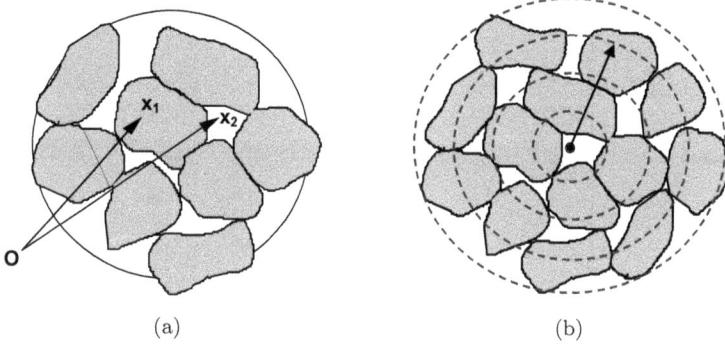

(a)   (b)

Figure 1.2.3. (a) Porosity at x_1 is 0, and at x_2 is 1, illustrating that porosity must not be defined at a point, but over a volume. (b) $\phi(R)$ can be defined as an average over a region of radius R (length of the arrow).

At the other extreme of length-scale, all porous rocks are heterogeneous on the scale of the pore size. Consider Fig. 1.2.3(a), where x_1 and x_2 are two locations in the rock. If we ask "what is the porosity at location x_1"?, then the answer would be "0", because point x_1 lies in a sand grain. On the other hand, point x_2 lies in a pore, so, strictly speaking, the porosity at x_2 is 1. Clearly, it makes no sense to talk about the porosity at some infinitely small mathematical point in the reservoir. When we discuss the porosity, we implicitly are referring to the *average* porosity in some small region.

Imagine that we could measure the porosity in a small spherical region of rock, of radius R, surrounding the point x_2, as in Fig. 1.2.3(b). Let's denote this average value as $\phi(R)$. For very small values of R, $\phi(R)$ would be 1. As R gets larger, the spherical region will start to encompass some of the nearby sand grains, and so $\phi(R)$ will decrease. In a typical sandstone, $\phi(R)$ will fluctuate with R, but eventually stabilise to some constant value. Eventually, as R gets large enough that the region crosses over into the next rock type, $\phi(R)$ might change abruptly. Schematically, we can represent this situation as in Fig. 1.2.4 (see Bear, 1972).

The minimum value of R needed for the porosity to stabilise is known as the "representative elementary volume" (REV), or as the "representative volume element" (RVE). When we talk about

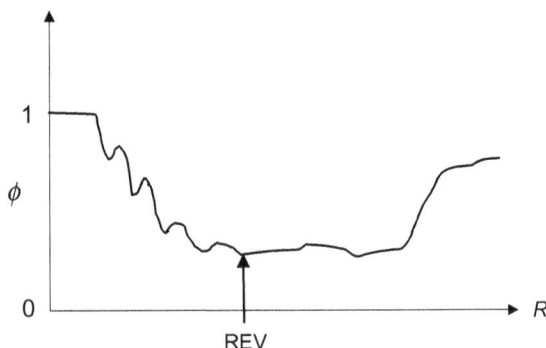

Figure 1.2.4 Porosity as a function of the size of the sampling region, showing the existence of an REV.

the porosity, we are usually implicitly referring to the porosity as defined on a length scale at least as large as the REV.

For sandstones with a uniform grain size distribution, the REV must be at least about ten grain diameters. However, for heterogeneous rocks such as many carbonates, the REV may be much larger, as heterogeneity may exist at many scales. In fact, there is no guarantee that a REV scale exists for a given rock. But in petroleum engineering, we always assume that a REV can be defined on the length scale of the gridblocks that are used in the numerical reservoir simulation codes. This will be discussed further in some later sections.

1.2.3.   *Saturation*

Reservoir rocks are never filled completely with oil, for reasons that will be discussed later in this chapter. Consider a rock that contains some oil and some water, as in Fig. 1.2.5. If the volume of water contained in a region of rock is V_w, the volume of oil is V_o, and the total volume of the pore space is V_p, then the *saturation* of each of these phases can be defined as the fraction of pore space that is occupied by that phase, i.e.

$$S_w = V_w/V_p, \quad S_o = V_o/V_p. \tag{1.2.3}$$

The saturation of each phase must, by definition, lie between 0 and 1. If oil and water are the only two phases present, then it is necessarily

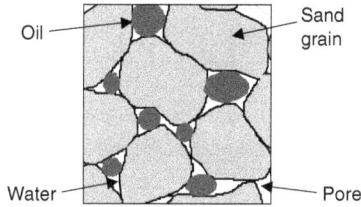

Figure 1.2.5. Schematic diagram of a porous rock containing oil and water.

the case that,

$$S_w + S_o = 1. \tag{1.2.4}$$

If there is also some hydrocarbon gas in the pore space, then

$$S_w + S_o + S_g = 1. \tag{1.2.5}$$

1.3. Permeability and Darcy's Law

1.3.1. *Darcy's Law*

The ability of a porous rock to transmit fluid is quantified by the property called *permeability*. Quantitatively, permeability is defined by the "law" that governs the flow of fluids through porous media — *Darcy's law*. This law was formulated by French civil engineer Henry Darcy in 1856 on the basis of his experiments on water filtration through sand beds. Darcy's law is the most important equation in petroleum engineering.

Imagine a fluid having viscosity μ, flowing through a horizontal tube of length L and cross-sectional area A, filled with a rock or sand. The fluid pressure at the inlet is P_i, and at the outlet is P_o, as in Fig. 1.3.1.

According to Darcy's law, the fluid will flow through the rock in the direction from higher pressure to lower pressure, and the volumetric flowrate of this fluid will be given by

$$Q = \frac{kA(P_i - P_o)}{\mu L}, \tag{1.3.1}$$

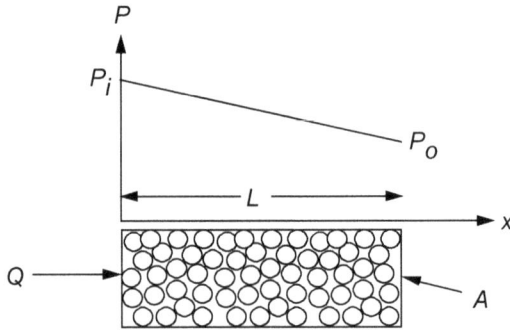

Figure 1.3.1. Experimental setup for measuring the permeability of a porous rock or sand.

where

$$Q = \text{volumetric flowrate, with units of m}^3/\text{s},$$
$$k = \textit{permeability} \text{ of the rock, with units of m}^2,$$
$$A = \text{cross-sectional area of the rock core, with units of m}^2,$$
$$P_i, P_o = \text{inlet/outlet pressures, with units of Pa,}$$
$$\mu = \text{viscosity of the fluid, with units of Pa·s,}$$
$$L = \text{length of the core, with units of m.}$$

This equation can be thought of as providing a definition of permeability, and it also shows us how to measure the permeability in the laboratory. This equation tells us that the flowrate is proportional to the area, inversely proportional to the fluid viscosity, and proportional to the pressure gradient, i.e. the pressure drop per unit length, $\Delta P/L$. Note that the permeability is a property of the rock; the influence of the fluid that is flowing through the rock is represented by the viscosity term in Darcy's law.

It is usually more convenient to work with the volumetric flow per unit area, $q = Q/A$. Darcy's law is therefore usually written as

$$q = \frac{Q}{A} = \frac{k(P_i - P_o)}{\mu L}, \tag{1.3.2}$$

where the flux q has dimensions of m/s. Please note that the flux is *not* the same as the velocity of the fluid particles,[1] and so it is perhaps easier to think of these units as m^3/m^2·s.

For the general case in which the flux may vary from point-to-point, we need a *differential* form of Darcy's law. The differential version of Eq. (1.3.2) for horizontal flow is,

$$q_x = \frac{-k}{\mu}\frac{dP}{dx}.$$ (1.3.3)

The minus sign is included to account for the fact that the fluid flows in the direction from *higher* to *lower* pressure.

For vertical flow, we must include a gravitational term in Darcy's law. To see why this is necessary, recall from fluid mechanics that if the fluid is stagnant, then the pressure distribution will be

$$P = P_o + \rho g z,$$ (1.3.4)

where z is the depth *below* some datum level, and P_o is the pressure at the datum level. So, there will be pressure gradient in a stagnant fluid, but there will be no flow. The "equilibrium pressure gradient" is, from Eq. (1.3.4),

$$\frac{dP}{dz}\bigg|_{\text{equilibrium}} = \rho g.$$ (1.3.5)

It seems reasonable to assume that fluid will flow through the rock only if the pressure gradient *exceeds* the equilibrium value given by Eq. (1.3.5). In this case, the actual driving force should

---

[1] *Note*: q is the flux based on the total nominal area of the core, but the fluid actually flows only through the pores, and not the grains! So, the total flux is given by $Q = qA$, but it can also be expressed as $Q = vA_{\text{pore}}$, where v is the actual mean velocity of the particles of fluid, and A_{pore} is area occupied by pores. Hence, $qA = vA_{\text{pore}}$, so $v = q(A/A_{\text{pore}}) = q/\phi$. For example, if q is 1 cm/h in a reservoir of 10% porosity, the actual mean velocity of the oil molecules as they travel through the pore space is 10 cm/h.

be $(dP/dz) - \rho g$. For vertical flow, we therefore modify Eq. (1.3.3) as follows:

$$q_z = \frac{-k}{\mu}\left[\frac{dP}{dz} - \rho g\right] = \frac{-k}{\mu}\frac{d(P - \rho g z)}{dz}. \qquad (1.3.6)$$

Actually, this form of the equation holds for horizontal flow also, because in this case we can say that

$$q_x = \frac{-k}{\mu}\frac{d(P - \rho g z)}{dx} = \frac{-k}{\mu}\frac{dP}{dx}, \qquad (1.3.7)$$

since $d(\rho g z)/dx = 0$.

A convenient way of simplifying the form of these equations is to write them in terms of the *fluid potential* Φ, defined by

$$\Phi = P - \rho g z, \qquad (1.3.8)$$

in which case flow in an arbitrary direction n can be described by[2]

$$q_n = \frac{-k}{\mu}\frac{d\Phi}{dn}. \qquad (1.3.9)$$

The above equations assume that the permeability is the same in all directions, but in most reservoirs the permeability in the horizontal plane, k_H, is different than the vertical permeability, k_V; typically, $k_H > k_V$. The permeabilities in different directions within the horizontal plane may also differ, but this difference is usually not as great as that between k_H and k_V. The property of having different permeabilities in different directions is known as *anisotropy*. For flow in an anisotropic rock, we must modify Darcy's law, as follows:

$$q_x = \frac{-k_H}{\mu}\frac{d\Phi}{dx}, \quad q_z = \frac{-k_v}{\mu}\frac{d\Phi}{dz}. \qquad (1.3.10)$$

---

[2] *Caution*: if the rock is anisotropic, then Eq. (1.3.9) does not hold in an *arbitrary* direction, even if we use an appropriate value of k (de Marsily, 1986). The correct version of Darcy's law for an anisotropic rock must be written in terms of the *permeability tensor*, which is a symmetric 3×3 matrix that has *six* independent components. However, this tensorial form of Darcy's law is not typically used in most reservoir engineering calculations.

Another way to think about why fluid flow is controlled by the gradient of $P - \rho g z$ is as follows. You may recall from undergraduate fluid mechanics that Bernoulli's equation, which essentially embodies the principle of "conservation of energy", contains the terms

$$\frac{P}{\rho} - gz + \frac{v^2}{2} = \frac{1}{\rho} \left(P - \rho g z + \frac{\rho v^2}{2} \right), \qquad (1.3.11)$$

where P/ρ is related to the enthalpy per unit mass, gz is the gravitational potential energy per unit mass, and $v^2/2$ is the kinetic energy per unit mass.

Fluid velocities in a reservoir are usually very small, so the kinetic energy term is negligible, in which case the combination $P - \rho g z$ represents the "Bernoulli energy". It seems reasonable that the fluid would flow from regions of higher to lower energy, and, therefore, the driving force for flow should be the *gradient* (i.e. the rate of spatial change) of the quantity $P - \rho g z$.

These considerations also warn us that we should not expect Darcy's law to hold in cases where the kinetic energy term is not negligible. In fact, at high flowrates, we must modify Darcy's law by incorporating a quadratic term q^2 on the left-hand side of, say, Eq. (1.3.3). The resulting more general equation, called the Forchheimer equation, is necessary in some situations, such as in some gas reservoirs, and particularly near the wellbore, where the velocities are higher (Bear, 1972). However, Darcy's law is adequate in the vast majority of situations.

1.3.2. *Units of Permeability*

Permeability has dimensions of "area", so in the SI system it has units of m^2. However, in most areas of engineering it is conventional to use a unit called the "Darcy", which is defined by

$$1 \text{ Darcy} = 0.987 \times 10^{-12} \, m^2 \approx 10^{-12} \, m^2. \qquad (1.3.12)$$

The Darcy unit is defined such that a rock having a permeability of 1 Darcy would transmit $1 \, cm^3$ of water (which has a viscosity of

Table 1.3.1. Typical ranges of the permeability of various rocks and sands.

Rock type	k (D)	k (m^2)
Coarse gravel	10^3–10^4	10^{-9}–10^{-8}
Sands gravels	10^0–10^3	10^{-12}–10^{-9}
Fine sand, silt	10^{-4}–10^0	10^{-16}–10^{-12}
Clay, shales	10^{-9}–10^{-6}	10^{-21}–10^{-18}
Limestones	10^{-4}–10^0	10^{-16}–10^{-12}
Sandstones	10^{-5}–10^1	10^{-17}–10^{-11}
Weathered chalk	10^0–10^2	10^{-12}–10^{-10}
Unweathered chalk	10^{-9}–10^{-1}	10^{-21}–10^{-13}
Granite, gneiss	10^{-8}–10^{-4}	10^{-20}–10^{-16}

1 centipoise/s) through a region that has a cross-sectional area of 1 cm^2, if the pressure drop along the direction of flow was 1 atm/cm.

This definition is strange, in that it utilises different systems of units. Some people apply Darcy's law by first converting flowrates to cm^3/s, converting areas to cm^2, etc., in which case you must use the value of k in Darcies. Another method is to first convert all parameters to SI units, in which case you must then use the value of k in units of m^2 when you apply Darcy's law.

Typical ranges of the permeability of various rocks and sands are given in Table 1.3.1.

Table 1.3.1 shows that the permeability of geological media varies over many orders of magnitude. However, most reservoir rocks have permeabilities in the range of 0.1 mD to 10 D, and usually in the much narrower range of 10–1000 mD. Methods for measuring permeability in the laboratory are discussed in the chapter on core analysis.

1.3.3. *Relationship between Permeability and Pore Size*

The permeability depends on the porosity of the rock, and also on the *pore size*. Many models have been developed to try to relate the permeability to porosity, pore size, and other attributes of the pore space. The simplest model assumes that the pores are all circular tubes of the same diameter. Consider a set of circular pore tubes,

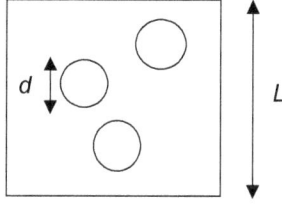

Figure 1.3.2. Idealised pore structure used to derive a relationship between permeability, porosity and pore size.

each of diameter d, passing through a cubical rock specimen of side L, as in Fig. 1.3.2, with a pressure difference ΔP imposed across the two parallel faces (on the page and behind the page) of the rock.

According to Poiseuille's equation for pipe flow (Dullien, 1992), the flow through each tube is given by

$$Q = \frac{\pi d^4}{128\mu} \frac{\Delta P}{L}. \tag{1.3.13}$$

If there are N such pores, the total flowrate will be

$$Q = \frac{N\pi d^4}{128\mu} \frac{\Delta P}{L}. \tag{1.3.14}$$

The total area of these pores, in the plane of the page, is $A_p = N\pi d^2/4$, and the porosity is $\phi = A_p/A = A_p/L^2$, where A is the macroscopic area normal to the flow. Hence, the total flowrate from Eq. (1.3.14) can be written as

$$Q = \frac{\phi d^2 A}{32\mu} \frac{\Delta P}{L}. \tag{1.3.15}$$

If we compare this flowrate with Darcy's law, $Q = kA\Delta P/\mu L$, we see that the permeability of this rock is $k = \phi d^2/32$. Lastly, we note that an isotropic rock should have only one third of its pores aligned in the x-direction, one third in the y-direction, etc. So, the permeability of this idealised porous rock is

$$k = \frac{\phi d^2}{96}. \tag{1.3.16}$$

A slightly more realistic model, in which the orientations of the pores are randomly distributed in 3D space, leads to exactly the same result (Scheidegger, 1974).

Equation (1.3.16) is often written in terms of the "specific surface", S/V, which is the total amount of surface area (S) per unit volume of rock; the result is

$$k = \frac{\phi^3}{6(S/V)^2}. \qquad (1.3.17)$$

This is often called the *Kozeny–Carman* equation (Bear, 1972). One justification for this equation is that the permeability is the inverse of the "hydraulic resistivity", and it is plausible that the resistance to flow, which is essentially due to viscous drag of the fluid against the pore walls, would be related to the amount of surface area of the pores.

In some versions of the Kozeny–Carman equation, the numerical factor six is replaced by another constant called the "tortuosity", τ. The tortuosity is sometimes claimed to represent the ratio of the actual fluid flow path from the inlet to the outlet, to the nominal fluid flow path L, but this is not true, and it is preferable to think of it as nothing more than an empirical fitting factor.

There have been many attempts to try to improve upon the Kozeny–Carman equation, by incorporating more information about the distribution of pore sizes, interconnectedness of the pores, etc. For the purposes of this chapter, it is sufficient to understand that the permeability is proportional to the square of the mean pore diameter.

1.3.4. *Permeability of Layered Rocks*

Most reservoir rocks are *layered*, with each layer having a different permeability. If fluid flows through a layered rock, either in the vertical direction (perpendicular to the layering) or the horizontal direction (parallel to the layering), it is possible to define an *effective permeability* that will allow us to treat the rock as if it were homogeneous, and use Darcy's law in its usual form.

For example, consider horizontal flow through a rock composed of N layers, each having permeability k_i and thickness H_i, as in

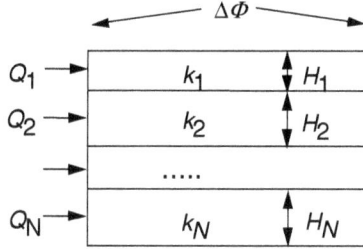

Figure 1.3.3. Fluid flow parallel to the layering of a layered rock.

Fig. 1.3.3. Within each layer, fluid will flow horizontally, according to Darcy's law:

$$Q_i = \frac{-k_i(H_i w)}{\mu} \frac{\Delta \Phi}{\Delta x}, \qquad (1.3.18)$$

where w is the thickness into the page.

The total flowrate is found by summing up the flowrates through each layer:

$$Q = \sum_{i=1}^{N} Q_i = \sum_{i=1}^{N} \frac{-k_i(H_i w)}{\mu} \frac{\Delta \Phi}{\Delta x} = \frac{-w}{\mu} \frac{\Delta \Phi}{\Delta x} \sum_{i=1}^{N} k_i H_i, \qquad (1.3.19)$$

but if we were to treat the rock as if it were a homogeneous rock mass with an effective permeability k_{eff}, then we would write Darcy's law as

$$Q = \frac{-k_{\text{eff}}(H_{\text{total}} w)}{\mu} \frac{\Delta \Phi}{\Delta x} = \frac{-w}{\mu} \frac{\Delta \Phi}{\Delta x} k_{\text{eff}} \sum_{i=1}^{N} H_i. \qquad (1.3.20)$$

If we compare Eqs. (1.3.19) and (1.3.20), we see that the effective permeability of the layered rock is

$$k_{\text{eff}} = \sum_{i=1}^{N} k_i H_i \left/ \sum_{i=1}^{N} H_i = \frac{1}{H} \sum_{i=1}^{N} k_i H_i. \right. \qquad (1.3.21)$$

Hence, the effective permeability for flow *along the layering* is the *weighted arithmetic mean* of the individual permeabilities, weighted by the thickness of the layers.

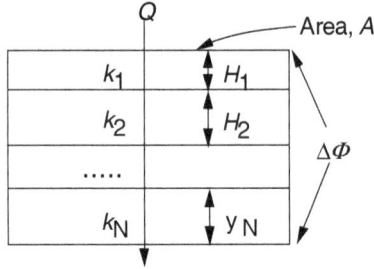

Figure 1.3.4. Fluid flow perpendicular to the layering through a layered rock.

Now imagine vertical flow through this layered system (Fig. 1.3.4). We again start by writing Darcy's law for each layer:

$$Q_i = \frac{-k_i A}{\mu} \frac{\Delta \Phi_i}{H_i}, \tag{1.3.22}$$

where A is the area normal to the flow direction, i.e. in the horizontal plane.

In the steady state, the flowrate through each layer must be the same, but the potential drop will be different. So, we put $Q_i = Q$ in each layer, and rewrite Eq. (1.3.22) in the form

$$\Delta \Phi_i = \frac{-\mu Q H_i}{A k_i}. \tag{1.3.23}$$

The total potential drop across all N layers is found by summing up the drops across each individual layer:

$$\Delta \Phi = \sum_{i=1}^{N} \Delta \Phi_i = \sum_{i=1}^{N} \frac{-\mu Q H_i}{A k_i} = \frac{-\mu Q}{A} \sum_{i=1}^{N} \frac{H_i}{k_i}. \tag{1.3.24}$$

The potential drop across an "equivalent" homogeneous rock of total vertical thickness H and area A would be

$$\Delta \Phi = \frac{-\mu Q H}{A k_{\text{eff}}} = \frac{-\mu Q}{A k_{\text{eff}}} \sum_{i=1}^{N} H_i. \tag{1.3.25}$$

Comparison of Eqs. (1.3.24) and (1.3.25) shows that

$$k_{\text{eff}} = \sum_{i=1}^{N} H_i \bigg/ \sum_{i=1}^{N} \frac{H_i}{k_i} = \left[\frac{1}{H} \sum_{i=1}^{N} \frac{H_i}{k_i} \right]^{-1}. \tag{1.3.26}$$

The expression on the right side of Eq. (1.3.26) is called the *weighted harmonic mean* of the permeabilities.

Equations (1.3.21) and (1.3.26) are similar to the equations for the overall conductivity of electrical resistors in parallel or series. However, this analogy can easily be remembered incorrectly, because the thickness also appears in these equations, and it appears in a different way in the two cases. Rather than trying to remember the analogy between electrical circuits and flow through layered rocks, it is safer to derive the laws for the effective permeability from first principles (or to refer to these notes).

Roughly speaking, the effective permeability for flow parallel to the layering is controlled by the permeability of the most permeable layer, whereas for flow transverse to the layering, the least permeable layer plays the controlling role.

1.3.5. *Permeability Heterogeneity*

In many reservoirs the heterogeneity is more complex than the simple layering shown in Figs. 1.3.3 and 1.3.4. Moreover, if oil is flowing towards a well, the flow geometry will be radial, and clearly the "series" and "parallel" models cannot be expected to apply.

In order to calculate fluid flow in a reservoir using either analytical or numerical methods (both of which will be covered later in this volume), it is necessary to replace the heterogeneous distribution of permeabilities with a single "effective" permeability. This difficult problem, which is known in petroleum engineering as "upscaling", will be covered in detail in a later volume.

For now, we only mention that the *geometric mean* often provides a good estimate of the effective permeability, for cases in which the heterogeneity is "randomly" distributed. The geometric mean of a permeability distribution is defined such that the natural logarithm of the geometric mean is the volumetrically-weighted average of the logarithm of the individual permeabilities. For example, if we have N different regions, "randomly" arranged, each with permeability k_i and volume fraction c_i, then the geometric mean is defined as

$$\ln(k_G) = \sum_{i=1}^{N} c_i \ln(k_i) = \sum_{i=1}^{N} \ln(k_i^{c_i}) = \ln \prod_{i=1}^{N} k_i^{c_i}, \qquad (1.3.27)$$

which is to say that

$$k_G = \prod_{i=1}^{N} k_i^{c_i} = k_i^{c_1} k_i^{c_2} k_i^{c_3} \cdots k_N^{c_N}. \qquad (1.3.28)$$

It can be proven that the geometric mean always lies between the arithmetic and harmonic means. Moreover, it can also be proven that, regardless of the precise geometric distribution of the local permeabilities, the effective permeability will always lie between the arithmetic and harmonic mean values (Beran, 1968). The fact that both the effective permeability and the geometric mean permeability are bounded by the arithmetic and harmonic means provides some justification for using the geometric mean as an approximation to the effective permeability, in cases in which the heterogeneity is "randomly" distributed.

1.4. Surface Tension, Wettability and Capillarity

The pore space of a reservoir rock always contains a mixture of different fluids. The manner in which these fluids distribute themselves within the pore space depends on the physico-chemical interactions between the various fluids and rock minerals. We now discuss some of the concepts and definitions needed to understand the distribution of fluids in the pore space.

1.4.1. *Surface Tension*

Consider an interface between two fluids, which for concreteness we take to be a gas and a liquid, as shown in Fig. 1.4.1. Now consider a molecule within the liquid, such as molecule A on the right. This molecule has a certain internal energy, u. There is a force of attraction between this molecule and all adjacent liquid molecules; call it F_{LL}.

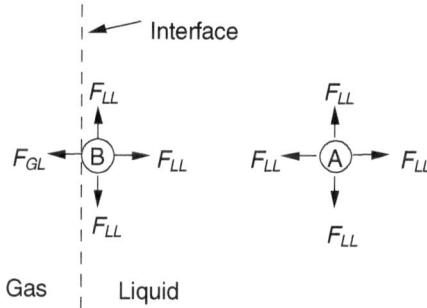

Figure 1.4.1. Simplified sketch of the forces acting on a fluid molecule.

The force of attraction between a liquid molecule and a gas molecule will be called F_{GL}.

Now imagine that we slowly pull molecule A towards the interface. Initially, it will be attracted equally to all of its neighbouring liquid molecules, and the net force on it will be zero. As it nears the interface, the liquid molecule to its right will be pulling on it with force F_{LL}, while the gas molecule to its left will be pulling with force F_{GL}. For concreteness, assume that $F_{GL} < F_{LL}$. In this case, there will be a net rightward force acting on molecule A, due to its neighbouring molecules. If we want to pull A to the surface, we must exert a leftwards force on it, and thereby do work on it. It follows that when molecule A reaches the surface, it will have *greater* energy than it did when it was in the bulk liquid.

The *total* excess energy that the liquid has due to its interface will obviously be proportional to the number of molecules at the surface, which is to say it will be proportional to the *area* of the interface. We can modify the usual thermodynamic expression for the internal energy to include surface energy as follows:

$$U = TS - PV + \gamma A, \tag{1.4.1}$$

where U is the internal energy, T is the temperature, S is the entropy, P is the pressure, V is the volume, A is the interface area, and γ is the "surface tension" between the liquid and the gas. If we treat the extensive variables (S, V, A) as the independent variables, then the

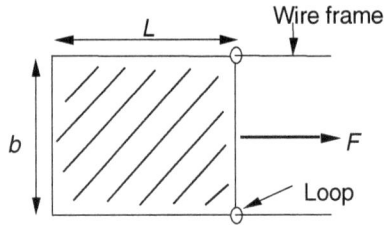

Figure 1.4.2. Thought experiment used to illustrate the significance of surface tension.

differential of Eq. (1.4.1) is

$$dU = TdS - PdV + \gamma dA. \tag{1.4.2}$$

Imagine now that we have a thin film of liquid in a device such as shown in Fig. 1.4.2. If we pull slowly on the slidable vertical bar with a force F, and move this bar by a small distance dL, then the work done by the external force on the liquid film will be FdL. By the first law of thermodynamics, the work done must equal the change in internal energy, so

$$dU = dW = FdL. \tag{1.4.3}$$

If we pull on the bar slowly and adiabatically (i.e. "reversibly"), the entropy change of the liquid film will be zero. Furthermore, the volume of the film is negligible, so PdV will be essentially zero. So, by Eq. (1.4.2),

$$dU = \gamma dA. \tag{1.4.4}$$

However, $A = bL$, and so $dA = bdL$, and Eq. (1.4.4) then gives

$$dU = \gamma bdL. \tag{1.4.5}$$

Equating Eqs. (1.4.3) and (1.4.5) shows that

$$FdL = \gamma bdL \Rightarrow F = \gamma b. \tag{1.4.6}$$

In other words, the external effect of surface tension is the same as if the interface were exerting a force of magnitude γ per unit length of the edge of the interface. Because of this interpretation, it is often convenient to treat an interface like an elastic membrane that exerts a force along its perimeter. Equation (1.4.6) also implies that γ has dimensions of force/length, and so it has SI units of N/m. Typical values for an oil/water interface are 0.01–0.05 N/m.

1.4.2. *Capillary Pressure*

Because of surface tension, the pressures within two fluid phases that are in mechanical equilibrium with each other across a curved interface will *not* be equal. To prove this, consider a bubble of gas, of radius R, inside a liquid that is contained in a rigid, thermally insulated container, as shown in Fig. 1.4.3. The pressure in the gas is P_G, the pressure in the liquid is P_L, and the surface tension of the gas–liquid interface is γ.

According to Eq. (1.4.2), the total differential of the internal energy of this liquid + gas + interface system will be

$$dU = TdS - P_L dV_L - P_G dV_G + \gamma dA. \qquad (1.4.7)$$

Note, that we count the "volumetric" term for both the liquid and the gas, but we must count the interface only *once*. However, $V_G + V_L = V_{\text{container}} = \text{constant}$, so $dV_L = -dV_G$; hence

$$dU = TdS + P_L dV_G - P_G dV_G + \gamma dA. \qquad (1.4.8)$$

Now assume that the bubble grows slowly. Since the container is rigid and thermally insulated, $dS = 0$ and $dU = 0$ (i.e. no heat is

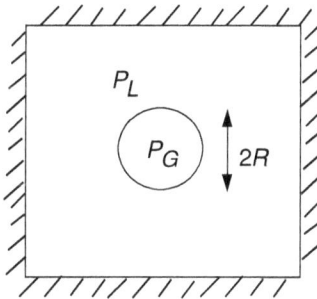

Figure 1.4.3. Thought experiment of a gas bubble surrounded by liquid, encased in a rigid, thermally insulated container, used to derive the Young–Laplace equation.

added to the system, and no work is done on the system, so the total entropy and energy remain the same). Hence,

$$\gamma \, dA = P_G \, dV_G - P_L \, dV_G = (P_G - P_L) \, dV_G. \qquad (1.4.9)$$

But $A = 4\pi R^2$, so $dA = 8\pi R \, dR$, and $V = 4\pi R^3/3$, so $dV = 4\pi R^2 \, dR$. Hence, Eq. (1.4.9) can be written as

$$8\pi\gamma R \, dR = (P_G - P_L) 4\pi R^2 \, dR, \qquad (1.4.10)$$

which is equivalent to

$$P_G - P_L = 2\gamma/R. \qquad (1.4.11)$$

This is the famous Young–Laplace equation, which states that the pressure inside the bubble is greater than the pressure outside, by an amount that is directly proportional to the surface tension between the two fluids, and *inversely* proportional to the radius of the bubble.

This pressure difference is known as the *capillary pressure*, i.e.

$$P_G - P_L \equiv P_{\text{cap}} = 2\gamma/R. \qquad (1.4.12)$$

A capillary pressure difference exists when any two fluids are in contact, not necessarily a liquid and a gas. For example, if we have a bubble of oil surrounded by water, then Eq. (1.4.12) would hold, with the subscripts G and L replaced by o for oil and w for water.

In a rock that is filled with oil and water, the interface between these two phases will be locally curved, and the pressures in the water and the oil phases will differ by an amount given by Eq. (1.4.12), where we define R to be the mean radius of curvature of the interface. Because of the inverse dependence of capillary pressure on radius, capillary pressures are more important in rocks with smaller pores than for rocks having larger pores.

1.4.3. *Contact Angles*

We now consider what happens when two fluids are in contact with a solid surface, as in Fig. 1.4.4, where a drop of liquid is sitting on a solid surface, surrounded by gas.

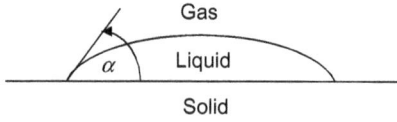

Figure 1.4.4. A drop of liquid sitting on a solid surface, surrounded by a gaseous phase.

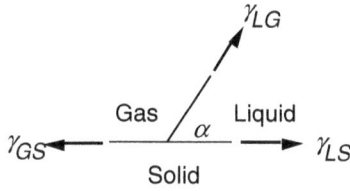

Figure 1.4.5. Force balance on the contact between three phases (liquid sitting on a solid surface, surrounded by a gaseous phase).

The slanted line is the tangent to the gas–liquid interface at the point where the interface meets the solid surface. The angle between the solid surface and the tangent, measured by rotating the solid surface towards the tangent, passing through the liquid (by definition the angle is measured through the denser phase), is called the *contact angle*, α.

We now do a free-body diagram (Fig. 1.4.5) and force-balance on the region where the three phases (solid, liquid, gas) meet; this is exactly like the "method of joints" used in analysing structural trusses. When we "slice" through each interface, the part of the interface that is "removed" will exert a tension γH on the "joint", where H is the distance into the page. For each surface tension, we will use subscripts to denote the two fluids that form the interface; i.e. γ_{LS} is the surface tension of the liquid–solid interface, etc. A force-balance in the horizontal direction yields

$$\Sigma F_{\text{horizontal}} = \gamma_{LS} - \gamma_{GS} + \gamma_{LG} \cos \alpha = 0,$$

$$\Rightarrow \cos \alpha = (\gamma_{GS} - \gamma_{LS})/\gamma_{LG}. \tag{1.4.13}$$

Several different cases can arise, depending on the relative magnitudes of the three surface tensions.

Case 1: $0 < \gamma_{GS} - \gamma_{LS} < \gamma_{LG}$:

In this case, the interfacial energy of a gas–solid interface is greater than that of a liquid–solid interface, so, roughly speaking, the solid will "prefer" to be in contact with the liquid. The right-hand side of Eq. (1.4.13) will lie between 0 and 1, so α will lie in the range

$$0 < \alpha < 90°. \tag{1.4.14}$$

In this case, we say that the liquid "wets" the solid surface, and the surface is called "water-wet" (although a better name would be "water-wettable", since a "water-wet" surface can be completely dry!).

Case 2: $\gamma_{LG} < \gamma_{GS} - \gamma_{LS} < 0$:

In this case, the interfacial energy of a gas–solid interface is less than that of the liquid–solid interface, so the solid will "prefer" to be in contact with the gas. The right-hand side of Eq. (1.4.13) will lie between -1 and 0, and so α will lie in the range

$$90° < \alpha < 180°. \tag{1.4.15}$$

In this case, we say that the liquid does not wet the surface; it will sit on the surface in a bubble-shape, with very little interfacial contact between the liquid and the solid surface; see Fig. 1.4.6.

Case 3: $|\gamma_{GS} - \gamma_{LS}| > \gamma_{LG}$:

In this case, the right-hand side of Eq. (1.4.13) does not lie between -1 and $+1$, and so there is no value of α that will satisfy the equation of mechanical equilibrium! To see what will happen in this case, consider the limiting case in which $\gamma_{GS} - \gamma_{LS} = \gamma_{LG}$, in which case

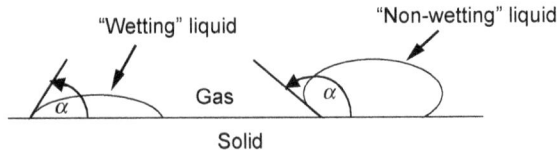

Figure 1.4.6. A wetting liquid (left) and a non-wetting liquid (right) on a solid surface.

$\cos\alpha = 1$, and $\alpha = 0°$. In this case, the liquid will spread out over the surface, creating as much liquid–solid interfacial area as possible (example: oil on water!). If $\gamma_{GS} - \gamma_{LS} > \gamma_{LG}$, then the same situation will occur, and the liquid will continue to flow until it forms a thin layer on the solid surface. By a similar argument, if $\gamma_{GS} - \gamma_{LS} < -\gamma_{LG}$, then the gas will spread out to cover as much of the solid surface as possible.

Most reservoir rocks are preferentially "water-wet" as opposed to "oil-wet". If a water-wet rock is partially saturated with oil and water, the pore walls will "prefer" to be in contact with water rather than with oil, and so the oil will tend to exist in the form of blobs, as in Fig. 1.4.6.

1.4.4. *Capillary Rise*

Consider a bucket containing some oil and some water, as in Fig. 1.4.7(a). These two fluids are not miscible, and oil is usually less dense than water, so the oil will sit on top of the water. The air on top of the oil is at atmospheric pressure.

Now imagine that we put a capillary tube, of radius R and composed of a water-wet material, in the bucket, as in Fig. 1.4.7(b). This tube can be thought of as a simple model of a porous rock. The

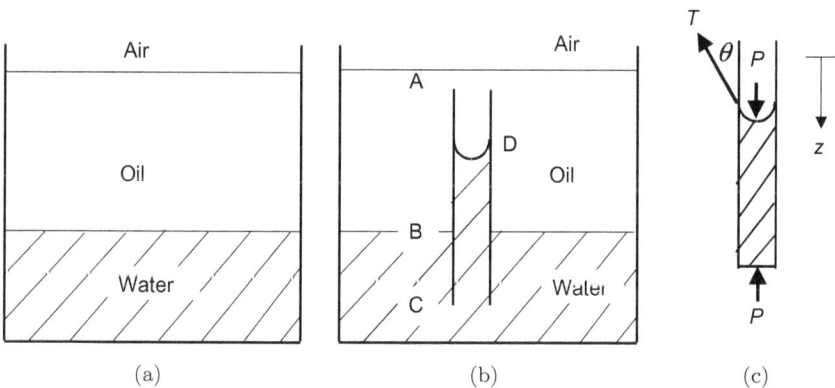

Figure 1.4.7. (a) Oil, air and water in a bucket; oil lighter than water. (b) If a capillary tube is inserted into the water, the water rises in the tube to a higher level. (c) Force balance on the column of water in the tube.

water will rise up in the tube by some height, h, above the original oil–water contact.

We now calculate the *capillary rise, h*, by performing a vertical force balance on the column of water in the tube, as in Fig. 1.4.7(c). The bottom of the column is pushed upwards by the pressure in the water at level C, acting over an area of πR^2. With the sign convention that the z-axis decreases with depth, this force is $-P_w(z_C)\pi R^2$. At the top of the column is a downwards-acting force due to the pressure in the oil at level z_D; this force is $+P_o(z_D)\pi R^2$. The surface tension exerts an upwards force along the entire wetted perimeter of the tube; its magnitude is $T = 2\pi\gamma_{ow}R$. It acts at an angle θ to the vertical, so its vertical component is $-2\pi\gamma_{ow}R\cos\theta$. Finally, gravity acts downwards on the column of water with a force

$$W = mg = \rho_w Vg = \rho_w\pi R^2(z_C - z_D)g. \qquad (1.4.16)$$

Summing all the vertical forces to zero gives

$$-P_w(z_C)\pi R^2 + P_o(z_D)\pi R^2 - 2\pi\gamma R\cos\theta + \rho_w\pi R^2(z_C - z_D)g = 0. \qquad (1.4.17)$$

The pressure in the oil at location D is equal to atmospheric pressure plus the pressure due to a column of oil of height $(z_D - z_A)$, i.e. $P_o(z_D) = P_{\text{atm}} + \rho_o g(z_D - z_A)$. Similarly, we can see that $P_w(z_C) = P_{\text{atm}} + \rho_o g(z_B - z_A) + \rho_w g(z_C - z_B)$. Inserting these expressions into Eq. (1.4.17) gives

$$-[P_{\text{atm}} + \rho_o g(z_B - z_A) + \rho_w g(z_C - z_B)]\pi R^2$$
$$+ [P_{\text{atm}} + \rho_o g(z_D - z_A)]\pi R^2$$
$$- 2\pi\gamma_{ow}R\cos\theta + \rho_w\pi R^2(z_C - z_D)g = 0, \qquad (1.4.18)$$

which can be solved to find the capillary rise, h:

$$z_B - z_D = h = \frac{2\gamma_{ow}\cos\theta}{(\rho_w - \rho_o)gR}. \qquad (1.4.19)$$

Hence, the height to which water would rise in a tube of radius R is proportional to the surface tension between the water and oil, and is inversely proportional to the radius of the tube.

We now examine the difference in pressures between the *oil at depth D* (outside the tube), and the *water at depth D* (inside the tube). By definition, this is the *capillary pressure* at depth D.

First, recall that $P_o(z_D) = P_{atm} + \rho_o g(z_D - z_A)$. Next, by starting at point A, going down to point B in the oil, and then going back up to point D in the water, we can find that $P_w(z_D) = P_{atm} + \rho_o g(z_B - z_A) - \rho_w g(z_B - z_D)$. Hence,

$$P_o(z_D) - P_w(z_D)$$
$$= P_{atm} + \rho_o g(z_D - z_A) - P_{atm} - \rho_o g(z_B - z_A) + \rho_w g(z_B - z_D)$$
$$= (\rho_w - \rho_o)g(z_B - z_D) = (\rho_w - \rho_o)gh = 2\gamma_{ow}\cos\theta/R. \quad (1.4.20)$$

In other words, $P_{cap} = 2\gamma_{ow}\cos\theta/R$. This is identical to the Young–Laplace equation that we derived earlier for an oil bubble in water, modified to account for the contact angle. Note also that the capillary pressure at any height h is equal to $(\rho_w - \rho_o)gh$. The theory we have just described is referred to as "capillary-gravity equilibrium".

1.4.5. *Oil–Water Transition Zone*

Most oils are less dense than water, so we might expect that a reservoir would contain only oil down to a certain depth, and only water below that depth, as would occur in a bucket containing oil and water. Although it is true that the rock is usually fully saturated with water below a certain level, on top of this zone is an *oil–water transition zone*, in which the water saturation decreases gradually with height. We can use the concept of capillary rise in a tube, along with a modified form of the parallel-tube model of a porous medium, to understand the existence of this oil–water transition zone in a reservoir.

We first return to our parallel tube model of a porous rock, but now imagine a *distribution* of different radii. Now, imagine that we place this porous rock into our bucket of oil and water:

According to Eq. (1.4.19), water would rise very slightly into a pore that has a large radius, but would rise very high in a small pore. If we have a distribution of pore radii, then at an elevation such as A in Fig. 1.4.8, all of the pores would be filled with water, and the

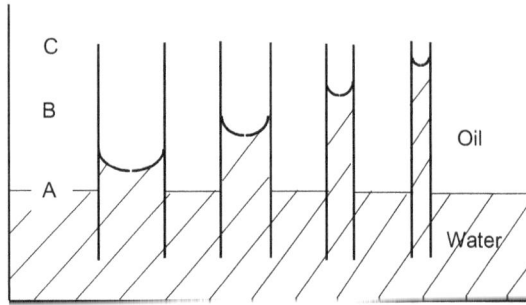

Figure 1.4.8. Similar to Fig. 1.16(b), but with a set of capillary tubes of different radii. According to Eq. (1.4.19), the water will rise higher in the smaller tubes.

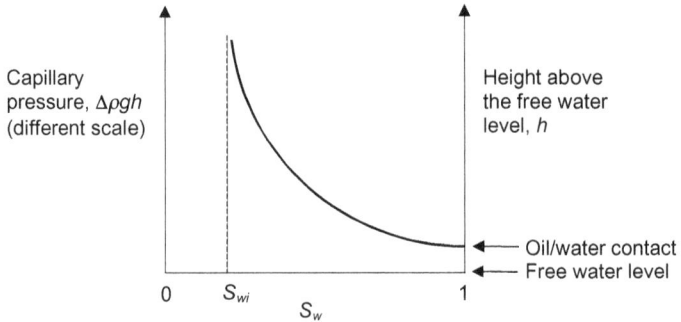

Figure 1.4.9. Capillary pressure (left scale) and height above the FWL (right scale), as functions of water saturation.

water saturation would be $S_w = 1$. At elevation B, some of the pores will be filled with water, and others with oil, so the water saturation will be $0 < S_w < 1$. Finally, at a high enough elevation, such as C, all the pores will be filled with oil, and $S_w = 0$.

Hence, the water saturation will be a decreasing function of the height h above the "free water level" (FWL), which is defined as the highest point in the reservoir where the capillary pressure is zero; see Fig. 1.4.9. According to our capillary tube model shown in Fig. 1.4.8, the water saturation will actually be equal to 1 up to some finite height above the FWL that is controlled by the radius of the largest pores. The highest point at which the saturation is equal to 1 is known as the "oil–water contact" (OWC). Note that in a water-wet reservoir, the OWC is above the FWL.

From Eq. (1.4.20), the "capillary pressure" is given by

$$P_{\text{cap}} = P_o(h) - P_w(h) = (\rho_w - \rho_o)gh, \qquad (1.4.21)$$

and so the y-axis in Fig. 1.4.9 essentially represents both the capillary pressure and the height above the FWL, which differ only by the multiplicative factor $(\rho_w - \rho_o)g$. Hence, this graph represents the water–oil transition zone, but also represents the capillary pressure function as a function of saturation.

Note that the relation $P_{\text{cap}} = (\rho_w - \rho_o)gh$ holds regardless of the specific rock geometry, assuming only that the rock is water-wet. However, the precise form of the $P_{\text{cap}}(S_w)$ curve shown in Fig. 1.4.9 depends on the pore geometry, and specifically on the pore-size distribution.

For the simple bundle-of-parallel-tubes model, one can derive an exact relationship between the $P_{\text{cap}}(S_w)$ curve and the pore-size distribution. For a real rock, in which the pores are interconnected, the relationship is not so simple, but it is always true that a narrow pore-size distribution corresponds to a P_c curve with a nearly horizontal shape, whereas a broader pore-size distribution yields a curve that increases more gradually, as shown in Fig. 1.4.10. The extreme case of a bundle of tubes in which all pores had the same

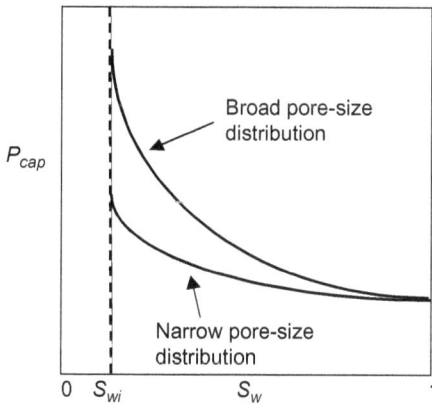

Figure 1.4.10. Capillary pressure curves for two rocks, with a narrow and a broad pore-size distribution, respectively.

radius R would yield a capillary pressure function that was essentially a horizontal line, at the value $P_{cap} = 2\gamma_{ow}\cos\theta/R$.

Note that whereas the bundle-of-tubes model predicts that the water saturation becomes zero at some (large) height above the FWL, in a real rock the water saturation never falls below some non-zero value S_{wi}, known as the *irreducible water saturation*, which is typically about 10% (Fig. 1.4.10). Hence, there is generally no region in a water-wet reservoir that contains only oil but no water!

1.4.6. *Leverett J-Function*

The capillary pressure function $P_{cap}(S_w)$ is often discussed in terms of a dimensionless function known as the *Leverett J-function*. We can see how this function arises by starting again with our bundle-of-tubes model.

Recall that for the simplest form of the bundle-of-tubes model, in which every tube has the same radius R, the capillary pressure is given by $P_{cap} = 2\gamma_{ow}\cos\theta/R$. In terms of the pore diameter, d, we can say that $P_{cap} = 4\gamma_{ow}\cos\theta/d$. But we also know that this bundle-of-tubes model predicts that $k = \phi d^2/96$. So, the pore diameter can be expressed as $d = (96k/\phi)^{1/2}$. If we plug this into our equation for P_{cap}, we find, after some rearrangement,

$$\frac{1}{\gamma\cos\theta}\sqrt{\frac{k}{\phi}}P_{cap} = \frac{1}{\sqrt{6}}. \qquad (1.4.22)$$

The left-hand side of Eq. (1.4.22) is essentially a dimensionless way of writing the capillary pressure function. For the bundle-of-uniform-tubes model, the right-hand side is a constant, but we know that this model is an oversimplification for real rocks. Moreover, we already saw above that capillary pressure in a rock varies with the saturation.

So, we can generalise Eq. (1.4.22) by replacing the constant on the right by some dimensionless function of saturation, which Leverett (1941) called the "*J-function*". This function will be a property of the rock, and its specific shape depends, in a complicated way, on the details of the pore geometry. We then express the capillary pressure in the form

$$\frac{1}{\gamma \cos \theta} \sqrt{\frac{k}{\phi}} P_{\text{cap}} = J(S_w). \tag{1.4.23}$$

The logic that underpins the use of the J-function is that, although properties such as k, ϕ and P_{cap} may vary throughout a sedimentary unit, it is generally true that the J-function, as defined by Eq. (1.4.23), is nearly invariant throughout the unit. Hence, if we measure P_{cap} for one core, and convert it into a J-function, we can then use Eq. (1.4.23) to estimate P_{cap} for other rocks in same unit.

Another use of the J-function is to take capillary pressure curves that are measured in the lab and convert them into capillary pressure curves for the reservoir. Assume that we measure P_{cap} in the lab using two fluids that have certain values of γ and θ, say γ_{lab} and θ_{lab}; then Eq. (1.4.23) takes the form

$$\frac{1}{\gamma_{\text{lab}} \cos \theta_{\text{lab}}} \sqrt{\frac{k}{\phi}} P_{\text{cap}}^{\text{lab}} = J(S_w). \tag{1.4.24}$$

In the reservoir, this rock would have the values of same k and ϕ, but the fluids would be different, so there will be different values of γ and θ, say γ_{res} and θ_{res}. Hence, in the reservoir we can say that

$$\frac{1}{\gamma_{\text{res}} \cos \theta_{\text{res}}} \sqrt{\frac{k}{\phi}} P_{\text{cap}}^{\text{res}} = J(S_w). \tag{1.4.25}$$

If we equate Eqs. (1.4.24) and (1.4.25), then we can rearrange and say that

$$P_{\text{cap}}^{\text{res}} = P_{\text{cap}}^{\text{lab}} \left(\frac{\gamma_{\text{res}} \cos \theta_{\text{res}}}{\gamma_{\text{lab}} \cos \theta_{\text{lab}}} \right). \tag{1.4.26}$$

Equation (1.4.26) shows how capillary pressures measured on a core in the lab can be converted to the values that would occur in the reservoir.

1.5.   Two-Phase Flow and Relative Permeability

1.5.1.   *Concept of Relative Permeability*

In Sec. 3, we defined and discussed the concept of permeability, in the context of a rock that is fully saturated with a single fluid phase. But we learned in Sec. 4 that reservoir rocks always contain at least two fluid phases, oil and water, and sometimes three phases, oil, water and gas. So, the concept of permeability must be extended to apply to situations in which more than one phase is present in the pore space.

The most obvious way to generalise Darcy's law to account for two-phase flow conditions is to assume that the flow of each phase is governed by Darcy's law, but with each parameter — pressure, viscosity and permeability itself — being specific to the phase in question, i.e.

$$q_w = \frac{-k_w}{\mu_w}\frac{dP_w}{dx}, \quad q_o = \frac{-k_o}{\mu_o}\frac{dP_o}{dx}, \tag{1.5.1}$$

where μ_w is the viscosity of water, k_w is the *effective permeability* of the rock to water, etc. The pressure in the oil phase and the water phase differ from each other by the capillary pressure, which is a function of the saturation.

However, it is more common to express the effective permeability of the rock to water as the product of the single-phase permeability, k (also known as the *absolute permeability*), and another parameter, k_{rw}, known as the *relative permeability of the rock to water*; likewise for oil. Note that the relative permeability function is dimensionless. Equation (1.5.1) can then be written as

$$q_w = \frac{-kk_{rw}}{\mu_w}\frac{dP_w}{dx}, \quad q_o = \frac{-kk_{ro}}{\mu_o}\frac{dP_o}{dx}. \tag{1.5.2}$$

The relative permeabilities of each phase are functions of the phase saturations. If part of the pore space is occupied by water, then the ability of oil to flow through the pore space will obviously be hindered, and *vice versa*. Hence, the relative permeability of a phase will be a monotonically increasing function of the saturation of that phase.

The precise shapes of these curves depend on the process that is occurring. Specifically, they have a different shape during *imbibition*, which is when a wetting phase displaces a non-wetting phase, than they have during *drainage*, which is when the non-wetting phase displaces the wetting phase.

We will first consider *imbibition*, such as occurs, for example, when we inject water into a water-wet reservoir to displace the oil. This process will start at $S_w = S_{iw}$, where S_{iw} is the irreducible water saturation, which is the water saturation that remained in the rock after oil had originally migrated into the reservoir. (The saturation $S_w = S_{iw}$ is also precisely the point at which the capillary pressure curve becomes unbounded.) Almost by definition, k_{rw} will be zero when $S_w = S_{iw}$. On the other hand, when $S_w = S_{iw}$, k_{ro} will be finite, but less than 1, as seen in Fig. 1.5.1.

Now imagine that we inject water into the rock, thereby increasing S_w. Since the relative permeability of a phase is an increasing function of the saturation of that phase, k_{rw} will increase, and k_{ro} will decrease. This imbibition process will continue until we reach a specific oil saturation, known as the *residual oil saturation*, $S_o = S_{or}$, at which k_{ro} has dropped to zero. This occurs at a finite value of the oil saturation, not at zero oil saturation, as one might have expected.

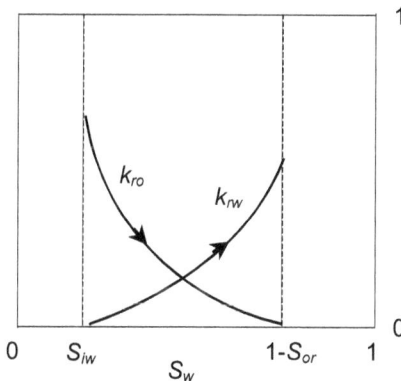

Figure 1.5.1. Relative permeability functions for water and oil.

These two values, S_{iw} and S_{or}, are also known as the relative permeability *end-points*, and the relative permeability values at these saturations are known as the *end-point relative permeabilities*.

The precise shapes of the relative permeability curves depend on the details of the pore structure of the rock. Power-law functions are often found to be useful in fitting these curves. Note that relative permeability curves are never linear functions of the saturation, although this simple linear form is sometimes assumed, particularly in fractured reservoirs, as data on the relative permeabilities of the fractures is rarely available. The range of validity of the assumption of linear relative permeability functions in fractured reservoirs has been investigated by de la Porte *et al.* (2005).

1.5.2.   *Irreducible Saturations*

The facts that S_{iw} was not equal to zero after primary drainage (when the oil first displaced the water to form the oil reservoir), and S_{or} is not zero after imbibition (when water is injected to flood the oil out of the reservoir) are of the utmost importance in reservoir engineering, but it is not obvious that these residual saturation values will not be zero. Also, in contrast to many other rock properties, which can, at least partially, be understood using the parallel-tube model of a porous rock, irreducible/residual saturations cannot be explained by the parallel-tube model. In fact, the parallel-tube model would erroneously predict that $S_{iw} = S_{or} = 0$. The phenomenon of irreducible/residual saturations is intimately related to the heterogeneity and interconnectedness of the pores in a rock.

We can gain a qualitative understanding of this phenomenon using the simplest pore-space model that incorporates some degree of heterogeneity and interconnectedness. Consider a *pore doublet*, as shown in Fig. 1.5.2, in which one pore branches off into two pores of different diameter, which then remerge to form a single pore:

Imagine that this doublet is initially filled with oil, as in Fig. 1.5.2(a). We now slowly inject water from the left. According to the Young–Laplace equation, Eq. (1.4.12), the capillary pressure in each pore is inversely proportional to the pore radius, i.e. it is

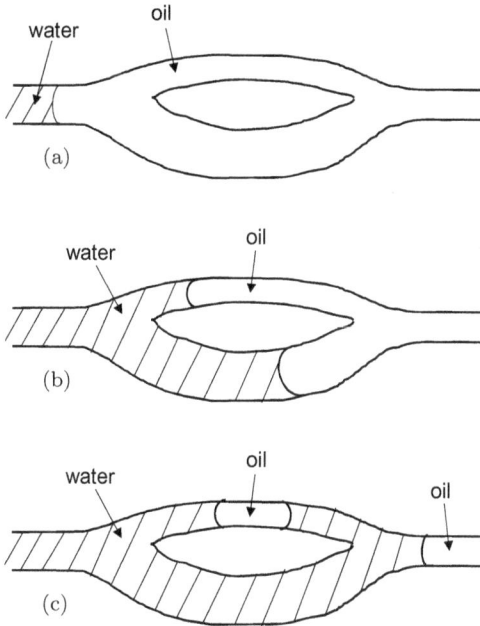

Figure 1.5.2. A pore doublet used to illustrate how small isolated blobs of oil can get trapped behind when water displaces oil in a reservoir, giving rise to a finite value of the residual oil saturation.

proportional to R^{-1}. However, according to Poiseuille's equation, Eq. (1.3.13), the "permeability" of a pore is proportional to R^2. So, from Darcy's law, Eq. (1.3.3), the mean velocity in either pore is proportional to $R^{-1} \times R^2 = R$. Hence, the water moves faster into the larger pore than into the smaller pore, as shown in Fig. 1.5.2(b).

When the water in the larger pore reaches the end of the doublet, it can enter the smaller pore from the far end, as in Fig. 1.5.2(c), thereby trapping some of the oil behind it. This is a simple demonstration of why the residual oil saturation, after the imbibition of water, is not equal to zero.

1.6. Electrical Resistivity

The electrical resistivity of a rock is not a property that directly affects oil production, nor does it appear in the governing equations of fluid flow in a reservoir, but it is nevertheless very important in reservoir engineering, because it can be measured *in situ* using *logging tools*, and its value can then be used to infer the value of the oil saturation. Practical issues related to the measurement of resistivity using logging tools, and the interpretation of these measurements, will be discussed in detail in the chapter on log analysis. In the present chapter, we will discuss some of the basic concepts and definitions related to the electrical resistivity of a fluid-saturated rock.

The flow of electrical current is governed by *Ohm's law*, which states that the current, I, flowing through any conductor, is equal to the voltage drop, ΔV, divided by the resistance, R:

$$I = \frac{\Delta V}{R}. \tag{1.6.1}$$

Electrical charge has units of coulombs, so current, which is the flow of charge, has units of coulombs/second. Resistance therefore has units of volt-seconds/coulomb, which are also known as *ohms*. In the form of Ohm's law given by Eq. (1.6.1), R will depend on the material properties, but also on the shape and size of the conductor.

Now consider a cylindrically shaped conductor, of length L and cross-sectional area A, as in Fig. 1.6.1. All other factors being equal, the current will be proportional to A, and inversely proportional to L.

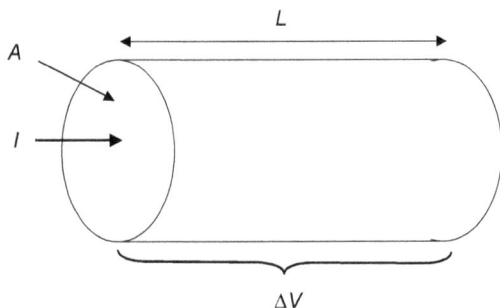

Figure 1.6.1. Cylindrical core with cross-sectional area A, length L, and axial voltage drop ΔV. If the material has conductivity σ, then the current will be given by $I = \sigma A \Delta V / L$.

So, we expect that the resistance R can be expressed as

$$R = \rho \frac{L}{A}, \tag{1.6.2}$$

where ρ, the *resistivity*, is an intrinsic property of the material, and does not depend on the geometry of the conductor. The resistivity has units of ohm-meters. Hence, Eq. (1.6.1) can be written as

$$I = \frac{A}{\rho} \frac{\Delta V}{L}. \tag{1.6.3}$$

We can also define the conductivity as $\sigma = 1/\rho$, in which case we can write Eq. (1.6.3) as

$$I = \sigma A \frac{\Delta V}{L}. \tag{1.6.4}$$

In this form, it is clear that Ohm's law is mathematically analogous to Darcy's law, with current (flow of electrical charge) being analogous to fluid flow, voltage drop analogous to pressure drop, and electrical conductivity analogous to the ratio of permeability/viscosity (i.e. the mobility).

The electrical conductivities of the minerals that typically form reservoir rocks are very low, as is the conductivity of hydrocarbon fluids. But the water that partially fills the pore space of reservoir rocks always contains salts such as NaCl or KCl, which render the water conductive. The conductivities of these *brines* are typically 10 orders of magnitude higher than that of the rock minerals. Hence, electrical current in a reservoir rock will flow mainly through that portion of the pore space that is occupied by water.

As with permeability and capillary pressure, we can get some idea of how the electrical conductivity depends on pore structure by appealing to the bundle of parallel tubes model. Consider such an idealised rock, as in Fig. 1.6.2, where A is the total area of the core, and the A_n are the areas of the individual pores.

Imagine that the pores are all filled with a brine of conductivity σ_w. If the core has length L into the page and the rock is subjected to a voltage drop ΔV along its length, then the current through the

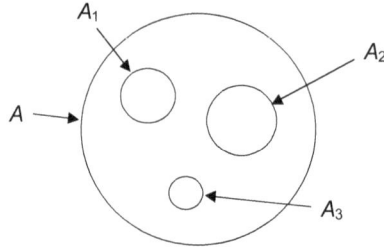

Figure 1.6.2. Cylindrical tube model of a porous rock, used in developing a simple model for the electrical formation factor.

nth tube will be

$$I_n = \sigma w \, A_n \frac{\Delta V}{L}.$$

(1.6.5)

The total current is the sum of the currents through each tube:

$$I = \sum_{n=1}^{N} I_n = \sum_{n=1}^{N} \sigma_w A_n \frac{\Delta V}{L} = \sigma_w \frac{\Delta V}{L} \sum_{n=1}^{N} A_n$$

$$= \sigma_w \frac{\Delta V}{L} A_{\text{pores}} = \sigma_w \frac{\Delta V}{L} \phi A.$$

(1.6.6)

If we compare this with Ohm's law, Eq. (1.6.4), we see that the effective conductance of the fluid-saturated rock is $\sigma_{\text{eff}} = \sigma_w \phi$. This quantity depends on the rock and on the brine. However, we are not really interested in the brine, so it would be preferable to extract out a parameter that reflects only the properties of the rock. We do this by defining — in general, independent of any pore geometry model — the *formation resistivity factor*, also known as the *formation factor*, as the ratio of the conductance of the brine to the effective conductance of the brine-saturated rock:

$$F \equiv \frac{\sigma(\text{brine})}{\sigma(\text{brine-saturated rock})}.$$

(1.6.7)

The resistivity is the inverse of the conductivity, so we can also say that

$$F \equiv \frac{\rho(\text{brine-saturated rock})}{\rho(\text{brine})}.$$

(1.6.8)

For our parallel-tube model, $\sigma_{\text{eff}} = \sigma_w \phi$, and so F is predicted to be equal to

$$F \equiv \frac{\sigma(\text{brine})}{\sigma(\text{brine-saturated rock})} = \frac{1}{\phi}. \qquad (1.6.9)$$

In contrast to the permeability, which has a strong dependence on pore size, the formation factor, according to the bundle-of-tubes model, has no dependence on pore size.

If we make the same argument as we did in the case of permeability, i.e. only one third of the pores are aligned in the direction of the voltage drop, then we would find

$$F = 3\phi^{-1}. \qquad (1.6.10)$$

This parallel-tube model correctly tells us that F will be larger for less porous rocks, but otherwise it is not accurate enough for engineering purposes. Experimental measurements of F tend to show a stronger dependence on porosity than the -1 power that appears in Eq. (1.6.10). Archie (1942) proposed generalising Eq. (1.6.10) by replacing both the factor of 3 and the exponent -1 with parameters that may vary from rock-to-rock. The result is the famous "Archie's law":

$$F = b\phi^{-m}. \qquad (1.6.11)$$

The parameter b is often called the *tortuosity*, and m is called the *cementation index*, but these names are outdated and not very useful.

Archie's law in the form of Eq. (1.6.11) can fit many sets of resistivity data consisting of different rocks from the same reservoir. For sandstones, the exponent m usually lies between 1.5 and 2.5, and is often close to 2; for carbonates, it can be as large as 4. The parameter b is usually close to 1.0. Figure 1.6.3 shows some data on Vosges and Fontainebleau sandstones, from Ruffet *et al.* (1991), fit with $b = 0.496$ and $m = 2.05$.

Although Archie's law is extremely useful in reservoir engineering, it should nevertheless be remembered that it is not a fundamental law of rock physics, but is merely a convenient curve-fit that is usually sufficiently accurate for engineering purposes.

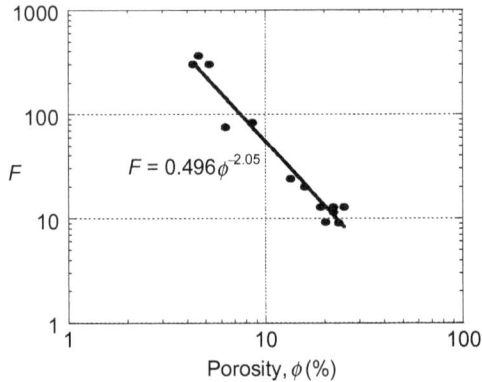

Figure 1.6.3. Measured values of formation factor on a set of Vosges and Fontainebleau sandstones, as a function of porosity (Ruffet *et al.*, 1991). Archie's law provides a reasonable fit, with $b = 0.496$ and $m = 2.05$.

It might appear that we could use Archie's law to estimate porosity, but there are much more accurate ways to estimate ϕ, as is explained in the modules on core and log analysis. The usefulness of Archie's law, and of resistivity measurements in general, is in estimating the water saturation. To understand how this is possible, we need to consider rocks that are partially saturated with water, and partially with oil.

For rocks that contain oil and water, we use the following generalisation of Archie's law, which is sometimes called *Archie's second law*:

$$F = b\phi^{-m}(S_w)^{-n}, \qquad (1.6.12)$$

where n is called the *saturation exponent*. In water-wet rocks, n is often close to 2, but in oil-wet rocks it may be as large as 10. If one assumes that n is constant throughout a reservoir, or at least throughout a certain rock unit, then Eq. (1.6.12) implies that electrical resistivity measurements in a borehole can yield estimates of the water saturation, and hence the oil saturation. This will be discussed in detail in the module on well logging.

The scenario described above is more complicated in shaly sands. In these rocks, the sand grains are coated by clay platelets. An ionic electrical double layer then builds up along the surface of these plates. This surface layer allows another path for current flow, separate from the current flow through the brine-saturated pores that were discussed above. To a good approximation, this surface current can be thought of as being in parallel with the pore-current, and so it adds an extra component to the conductivity of the rock. This extra conductivity depends on the electrochemical properties of the clays, but not on the intrinsic conductivity of the brine. The resulting generalisation of Eq. (1.6.11) for shaly sands is

$$\sigma(\text{brine-saturated rock}) = \frac{1}{F}(\sigma_w + BQ_v), \qquad (1.6.13)$$

where Q_v is the charge on the double layer, per unit volume, and B is a constant. Equation (1.6.13), which is called the Waxman–Smits equation, will be discussed further in the module on core analysis.

1.7. Fluid and Pore Compressibility

1.7.1. *Fluid Compressibility*

Quantities of oil are usually discussed in terms of barrels, which is a measurement of *volume*. Likewise, gas is often measured in terms of cubic feet. But the amount of volume taken up by a given mass of oil will depend on the pressure to which the oil is subjected. When we say that a well produces 100 barrels of oil a day, for example, we are measuring these barrels at "atmospheric pressure", which is 14.7 psi, or 101 kPa in SI units.

The relationship between the volume and pressure of a fluid is quantified in terms of the *fluid compressibility*, C_f, which is defined as the fractional derivative of volume with respect to pressure (at constant temperature, T, and for a fixed amount of mass):

$$C_f = -\frac{1}{V}\left(\frac{\partial V}{\partial P}\right)_T. \qquad (1.7.1)$$

Density is the inverse of specific volume, i.e. $\rho = 1/v$, where $v = V/\text{mass}$, so it follows from the chain rule of calculus that the compressibility can also be defined as

$$C_f = \frac{1}{\rho} \left(\frac{\partial \rho}{\partial P} \right)_T . \tag{1.7.2}$$

The dimensions of C_f are 1/pressure, and so the units are 1/psi or 1/Pa. The compressibility of a fluid usually varies with pressure, but typical values of C_f are 0.5×10^{-9}/psi for water and about 1.0×10^{-9}/psi for oil.[3]

1.7.2. *Pore Compressibility*

As oil flows from the reservoir to a well, two changes occur at any given location in the reservoir: the pressure in the oil decreases, and the pore space contains less oil. It is important to be able to relate the change in the amount of oil stored in the pore space of the reservoir to the change in the fluid (oil) pressure. This relation will obviously involve the fluid compressibility, but also involves a property of the rock known as the *pore compressibility.*

Consider a porous rock as shown below, with total (bulk) volume V_b, pore volume V_p, and mineral grain volume V_m. Imagine that the rock is compressed from the outside by a hydrostatic pressure P_c, called the *confining pressure.* Inside the pore space is a fluid at some *pore pressure,* P_p, which acts over the walls of the pores. The confining pressure tends to compress both the bulk rock and the pores, whereas the pore pressure tends to cause V_b and V_p to increase. (The confining pressure P_c, which acts on the rock, should not be confused with the capillary pressure P_{cap}, which acts within the fluid.)

Recall from Eq. (1.7.1) that for a homogeneous solid or liquid of volume V, subjected to a confining pressure P, the compressibility, C,

---

[3] *Note:* when oil is taken from its pressurised state in the reservoir up to the surface where it is at atmospheric pressure, any gas that had been dissolved in the oil will be released, and the oil will actually *shrink*! This phenomenon cannot be described in terms of the compressibility of liquid oil; this shrinkage must be taken into account when doing material balance calculations on a reservoir. However, Eqs. (1.7.1) and (1.7.2) are applicable to the incremental pressure changes that occur to the oil when it is flowing inside the reservoir.

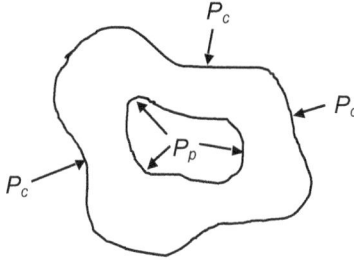

Figure 1.7.1. Schematic diagram of a porous rock, subjected to an external confining pressure P_c, and an internal pore pressure, P_p.

is defined as

$$C = \frac{-1}{V}\left(\frac{\partial V}{\partial P}\right). \tag{1.7.3}$$

For a porous rock (Fig. 1.7.1), we need to consider two volumes, the pore volume and the bulk volume, and two pressures, the pore pressure and the confining pressure. So, we can define four different compressibilities (Zimmerman, 1991):

$$C_{bc} = \frac{-1}{V_b}\left(\frac{\partial V_b}{\partial P_c}\right)_{P_p}, \quad C_{bp} = \frac{1}{V_b}\left(\frac{\partial V_b}{\partial P_p}\right)_{P_c}, \tag{1.7.4}$$

$$C_{pc} = \frac{-1}{V_p}\left(\frac{\partial V_p}{\partial P_c}\right)_{P_p}, \quad C_{pp} = \frac{1}{V_p}\left(\frac{\partial V_p}{\partial P_p}\right)_{P_c}. \tag{1.7.5}$$

The first subscript refers to the volume in question, either "bulk" or "pore", and the second subscript refers to the pressure that is varying, either "confining" or "pore". The pressures written outside the derivative indicates that this pressure is held constant.

The pore compressibility with respect to changes in pore pressure, C_{pp}, is useful in material balance calculations. The numerical sum of the fluid compressibility and the pore compressibility, which is known as the *total compressibility*, $C_t = C_f + C_{pp}$, appears in the pressure diffusivity equation that is used in well-test analysis. The bulk compressibility C_{bc} influences the velocity of seismic compressional waves. The bulk compressibility C_{bp} is relevant to subsidence calculations.

The numerical values of the various porous rock compressibilities are controlled, as are all petrophysical properties, by the geometry of the pore space. Roughly speaking, flat, crack-like pores are very compressible, whereas fatter pores in the shape of circular tubes are not very compressible.

The numerical values of the pore compressibility vary from one rock type to another. Furthermore, the pore compressibility often varies strongly with the pore pressure. Roughly, one can say that in a sandstone reservoir, the pore compressibility C_{pp} is on the order of about $1 - 10 \times 10^{-6}$/psi, or $2 - 15 \times 10^{-4}$/MPa. More details and numerical values can be found in *Compressibility of Sandstones* by Zimmerman (1991), and *The Rock Physics Handbook* by Mavko *et al.* (2009).

References

Archie, G. E. (1942). The electrical resistivity log as an aid in determining some reservoir characteristics, *Petrol. Trans. AIME*, **146**, 54–62.

Bear, J. (1972). *Dynamics of Fluids in Porous Media*, American Elsevier, New York.

Beran, M. (1968). *Statistical Continuum Theories*, Interscience, London.

de la Porte, J. J., Kossack, C. A. and Zimmerman, R. W. (2005). The effect of fracture relative permeabilities and capillary pressures on the numerical simulation of naturally fractured reservoirs. SPE Annual Technical Conference held in Dallas, (SPE 95241).

de Marsily, G. (1986). *Quantitative Hydrogeology*, Academic Press, San Diego.

Dullien, F. A. L. (1992). *Porous Media: Fluid Transport and Pore Structure*, 2nd ed., Academic Press, San Diego.

Leverett, M. C. (1941). Capillary behaviour in porous solids, *Petrol. Trans. AIME*, **142**, 159–172.

Mavko, G., Mukerji, T. and Dvorkin, J. (2009). *The Rock Physics Handbook*, 2nd ed., Cambridge University Press, Cambridge.

Ruffet, C., Gueguen, Y. and Darot, M. (1991). Complex conductivity measurements and fractal nature of porosity, *Geophysics*, **56**(6), 758–768.

Scheidegger, A. E. (1974). *The Physics of Flow through Porous Media*, University of Toronto Press, Toronto.

Zimmerman, R. W. (1991). *Compressibility of Sandstones*, Elsevier, Amsterdam.

Questions

1. Consider a reservoir that is shaped like a circular disk, 10 m thick, and with a 5 km radius in the horizontal plane. The mean porosity of the reservoir is 15%, the water saturation is 0.3, and the oil saturation is 0.7.

   (a) Ignoring the expansion of the oil that would occur when it is produced from the reservoir, how many barrels of oil are in this reservoir? One barrel = 0.1589 m^3.
   (b) If the density of the oil is 900 kg/m^3, how much oil (in kg) is contained in the reservoir?

2. In a laboratory experiment, a pressure drop of 100 kPa is imposed along a core that has length of 10 cm, and a radius of 2 cm. The permeability of the core is 200 mD, its porosity is 15%, and the viscosity of water is 0.001 Pa·s.

   (a) What will be the volumetric flowrate Q of the water, in m^3/s?
   (b) What is the numerical value of $q = Q/A$, in m/s?

3. Imagine that the rock in problem 2 can be represented by the parallel-tube model.

   (a) Estimate the mean pore diameter, d, using Eq. (1.3.16).
   (b) What is the mean velocity, v, of the water particles in the rock?
   (c) The importance of the inertia term, relative to the pressure term in, say, Eq. (1.3.11), can be quantified by the Reynolds number, defined as $Re = \rho v d/\mu$. What is the Reynolds number in this experiment? Note that Darcy's law is only accurate when $Re < 1$ (Bear, 1972).

4. Consider a layered reservoir consisting of alternating layers, 1 m thick, of rock 1, rock 2 and rock 3, where $k_1 = 1000$ mD, $k_2 = 100$ mD and $k_3 = 10$ mD.

   (a) What is the effective permeability of this rock, if fluid is flowing parallel to the layering?
   (b) What is the effective permeability of this rock, if fluid is flowing perpendicular to the layering?

(c) Imagine that the reservoir consists of these three rock types, in equal volumetric proportions, but occurring in a "random" spatial distribution. Estimate the effective permeability in this case.

5. Consider a small blob of oil surrounded by water. The surface tension between the oil and water is 0.02 N·m. If the radius of the blob is 0.05 mm, what is the value of the capillary pressure? Is the pressure higher in the oil or the water?

6. Consider again the parallel tube model of a rock. Assume that the diameter of each pore is $20\,\text{mm}$, $\gamma = 0.02\text{N·m}$, $\rho_o = 900\,\text{kg/m}^3$, $\rho_w = 1000\,\text{kg/m}^3$, the contact angle θ is $45°$, and $g = 9.8\,\text{m/s}^2$. If this rock is placed in a tank containing water overlain by oil, as in Fig. 1.4.7, to what height will the water rise in the pores?

7. Consider a homogeneous reservoir with $\phi = 0.20$, $k = 200\,\text{mD}$, water–oil surface tension of $\gamma = 0.03\,\text{N/m}$, and oil–water contact angle of $35°$. The oil density is $850\,\text{kg/m}^3$, and the water density is $1050\,\text{kg/m}^3$. In the lab, we determine that the irreducible water saturation occurs when the J-function is equal to 4.23. What will be the height of the oil–water transition zone in the reservoir?

Hint: Use Eq. (1.4.23) to convert $J(S_{wi})$ to P_{cap}, and use Eq. (1.4.21) on to convert P_{cap} to height.

Chapter 2

Introduction to Enhanced Recovery Processes for Conventional Oil Production

Samuel C. Krevor* and Ann H. Muggeridge[†]

Department of Earth Sciences and Engineering,
Imperial College London,
Kensington, London SW7 2AZ, UK
*s.krevor@imperial.ac.uk
[†]a.muggeridge@imperial.ac.uk

2.1. Introduction: Definition, Techniques and the Global Role of EOR

2.1.1. *The Aims of this Module*

This series of lectures is designed to provide an introduction and overview of enhanced oil recovery (EOR) technologies. It encompasses technologies applied today or at pilot scales in the past towards the recovery of conventional, medium to light, crude oil. The lectures will begin with a general overview of the role of EOR in oil production today and a summary of the most recent forecasts for the coming decades. Next, EOR will be described generally within the framework of the goal of enhancing the recovery factor by increasing microscopic and macroscopic displacement efficiencies. The third section provides some detail about the working of gas injection processes in the framework of fractional flow analysis to evaluate improvement in displacement efficiency. The final section includes a discussion of the more common chemical EOR techniques including the emerging low salinity flooding technology.

2.1.2.   *Definitions and Techniques*

EOR processes are oil recovery strategies that use the injection of fluids, chemicals and heat into a reservoir that alter the thermophysical or chemical properties of the multi-phase fluid–rock system. Table 2.1.1 shows broad categories of EOR technologies relative to primary and secondary recovery techniques. There is still some discussion about the precise definition (Hite *et al.*, 2003), but most find it useful to define EOR as those technologies that enhance production through either the injection of heat or fluids into the reservoir that were not initially present (Hite *et al.*, 2003; Lake, 1989; Muggeridge *et al.*, 2012). They are distinct from secondary water injection and other pressure maintenance techniques. They are often tertiary recovery processes, but there is no requirement that this be the case and indeed most EOR techniques are most effective if applied as secondary recovery mechanisms. The term improved oil recovery (IOR), can encompass EOR techniques but also includes reservoir management practices such as well placement

Table 2.1.1. EOR defined relative to other production techniques.

Primary production
- Natural lift
- Artificial lift

Secondary production
- Waterflooding
- Pressure maintenance

EOR
- Thermal stimulation
- Chemical EOR
- Gas injection processes

Improved oil recovery
- Includes EOR
- New wells
- Reservoir management to increase sweep

Unconventional
- Shale oil — light tight oil
- Extra heavy crude
- Kerogen (oil shale)

and injection/production scheduling that increases production by targeting bypassed regions of the reservoir. The term unconventional production refers to production of hydrocarbons from tight reservoirs where hydraulic fracturing is required.

A wide range of gas injection, chemical and thermal EOR techniques have been used on commercial scales since the 1970s. Their geographic deployment discussed in the next section reflects a combination of local economic, geologic and regulatory circumstances that favour a specific production technology. Table 2.1.2 shows screening criteria for the deployment of EOR technologies in the three broad categories. Gas injection can be applied across a wide range of reservoir permeability but it is favoured in deeper reservoirs where high pressures lead to miscibility with the resident hydrocarbon. Gas injection targets medium to light crudes and can be economic even when the remaining oil saturation is as low as 20%. Steam injection by comparison is applied to heavy, unconventional oil. It requires high oil saturations, reservoir permeabilities and deployment in shallower reservoirs to be economic. On the other hand, it usually targets heavy to extra heavy crudes that could not otherwise be produced at all. Chemical EOR is a term used to describe techniques that alter the water chemistry, either by adding artificial chemicals (polymers, surfactants) or changing the salt content of the injected water (low salinity or high sulphate). Polymers are typically used for viscous oils in cooler, high permeability sandstone reservoirs as long as the formation water is not too saline, although new polymers are being developed that can target more challenging environments (high salinity formation water, higher temperatures, carbonates).

Table 2.1.2. EOR Screening criteria (Taber *et al.*, 1997a, 1997b).

Method	API	Saturation	Depth (m)
Nitrogen	35–48	40–75	>1,800
Hydrocarbon	24–41	30–80	>1,200
CO_2	22–36	20–55	>800
Polymer	>15	50–80	<2,700
Steam	>8	40–66	<1,400

Low salinity water injection is only likely to be successful in sandstone reservoirs where the formation water has a high salinity, the oil contains polar compounds and there is a proportion of clay (kaolinite) in the rock.

2.1.3.   The Role of EOR in Current and Future Global Oil Production

In 2014, EOR processes contributed 1.7 million barrels of oil per day to global oil production (Koottungal, 2014; IEA, 2013b). This represents about 2% of global oil production and was dominated by production in four countries, the USA, Venezuela, Canada and Indonesia (Fig. 2.1.1). In these countries, production from EOR constitutes a larger fraction of the total relative to the world average. EOR made up 10% of total production in the USA since 1992, 10% of production in Canada and 16% in Venezuela in 2014.

The technologies applied for EOR are also distributed unevenly around the globe. Steam injection contributes just over half of all oil produced from EOR technologies whereas the injection of a gas, predominantly CO_2 or hydrocarbon fluids, contributes most of the remainder. Chemical EOR processes have remained little used

Global oil production from EOR in 2014
Total production: 1.7 million barrels per day

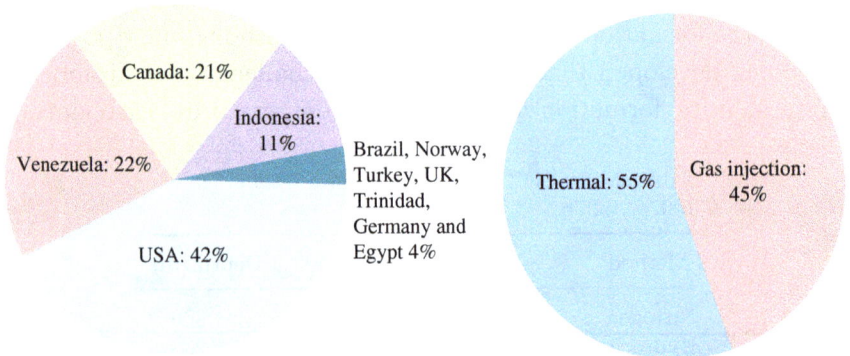

Figure 2.1.1. Oil produced using EOR techniques globally in 2014.
Source: Koottungal (2014).

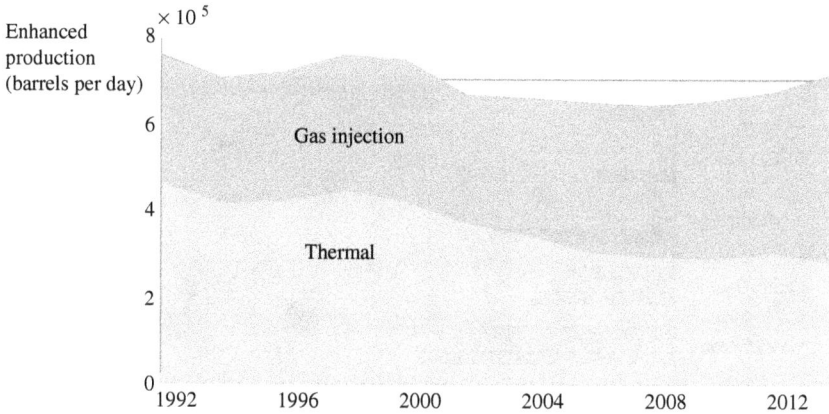

Figure 2.1.2. Oil produced using EOR techniques in the USA by technology since 1992.

Source: Koottungal (2014).

outside of China since the 1990s but have undergone a significant amount of research and development. In the USA about 60% of the oil from EOR came from gas injection and the remaining was produced using thermal methods, nearly all steam injection. Additionally, the fraction of oil recovered using gas injection has been steadily increasing (Fig. 2.1.2). In Canada, by contrast, over 80% of enhanced recovery is from steam injection and only 10% comes from gas injection. In Venezuela, about two-thirds of the enhanced production is from steam injection and the remaining third from hydrocarbon gas injection. The production in Indonesia from EOR is entirely from the largest steam flood project in the world, the Duri field. Steam injection makes up most of the remaining 4% of EOR projects around the world. The wide geographic deployment of steam injection is indicative of those locations, particularly in Canada and Venezuela, where there are significant resources of heavy crude oil. The relatively limited geographic distribution of gas injection is in large part due to the lack of large sources of CO_2 outside of the Rocky Mountain region of the Western USA.

Near term projections for oil produced through EOR show that it will continue to play an important but limited role in oil production through the middle of the 21st century. The most recent estimate

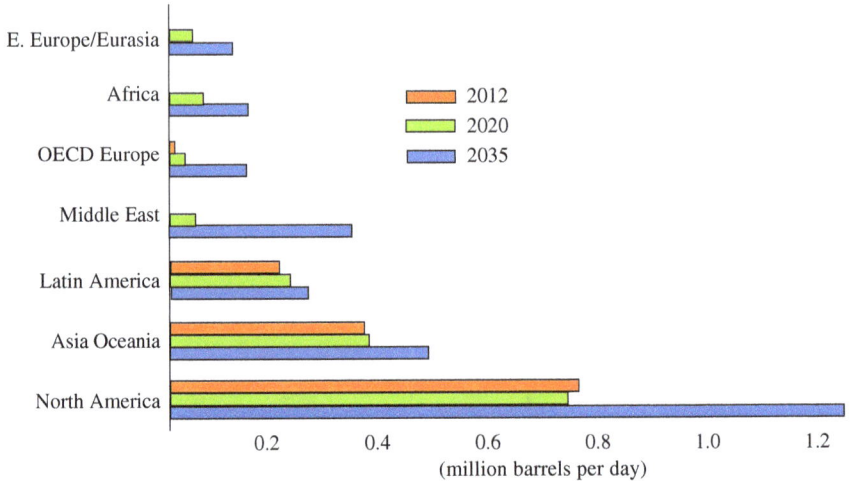

Figure 2.1.3. Modelled projections of oil production from all EOR operations globally.

Source: (IEA, 2013b).

by (IEA, 2013b) shown in Fig. 2.1.3 projects that production from EOR will increase only to 3.4 million barrels per day by 2035. In the USA, projections estimate a doubling or tripling of oil produced from CO_2 injection, from around 250,000 barrels per day today to somewhere between 450,000 and 1,000,000 barrels per day by 2040, depending primarily on the price of oil (USEIA, 2014; Wallace and Kuuskraa, 2014). Based on information about specific projects coming online before 2016 the study of (Wallace and Kuuskraa, 2014) predicts that much of this increase will take place prior to 2020. The limited application of EOR technologies projected is primarily due to the abundance of oil still recoverable through secondary production methods. However, it has also been impacted by the rapid development of techniques for producing unconventional oil and particularly shale oil, or so-called "light tight oil". Production from these reservoirs has a rapid rate of return, with payback on the investment in as little as 2 years and thus favours development by smaller companies with limited borrowing capacity (IEA, 2013a). Thus, while in the USA total production from EOR is projected to double slowly over 30 years, total oil production is expected to have nearly doubled from 4.5 to 9.6 million barrels per day in the decade 2010–2020. This is almost entirely due to production from the Bakken, Eagle Ford and Permian basin tight oil formations (USEIA, 2014). Thus in the USA, EOR will play a diminishing role over the next decade.

While EOR will be used to produce a small fraction of the petroleum supply as long as low cost production technologies keep up with demand, there is significant importance in the potential for the deployment of EOR to increase ultimately recovered resources. The average recovery factor for mature conventional crude projects ranges from 20% to 40%. The promise of EOR is to increase these recovery factors to the range 50–70% (Smalley *et al.*, 2009). An approximate understanding of the potential can be found as follows. Various estimates place remaining oil in place at around 10^{13} barrels. Currently an average recovery factor for fields is around $R_F = 0.25$. This suggests around 2.5×10^{12} barrels remain to be recovered without EOR. An increase in the recovery factor by 10% points, however, would result in 3.5×10^{12} barrels of oil to be recovered, an increment of 10^{12} barrels of oil, which is nearly as much as has been already produced historically! And a 50% increase in oil supply over our previous estimate.

Various global estimates using analyses ranging in complexity suggest that between 300 and 1,000 billion barrels of oil may be incrementally recovered through the application of EOR worldwide (ARI, 2009; IEA, 2013b; Wallace and Kuuskraa, 2014) (see Figs. 2.1.4 and 2.1.5). The US Energy Information Agency places oil supply meeting at least 25 years of demand. The widespread deployment of EOR extends this by at least 50% and thus it continues to be seen as a technology waiting in the wings but of major importance.

2.2.   Enhancing the Recovery Factor

2.2.1.   *The Recovery Factor*

The recovery factor, R_F, for a given oil field is the ratio of oil produced, N_P, to that estimated to be initially in place, N, both

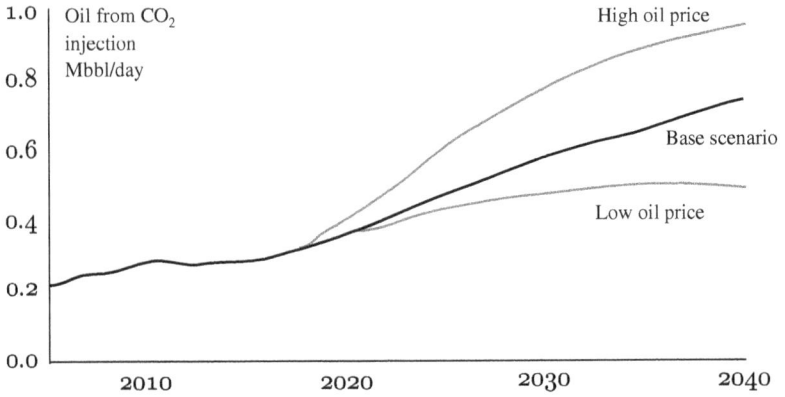

Figure 2.1.4. Modelled projections of oil produced from CO_2-EOR in the USA. *Source*: (USEIA, 2014).

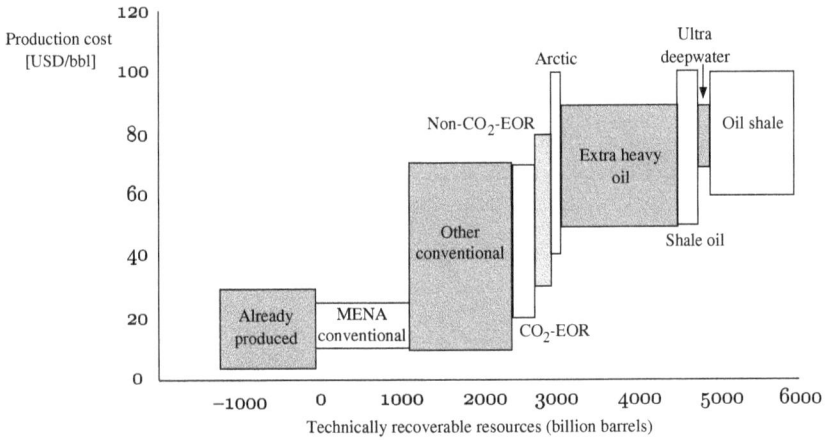

Figure 2.1.5. Technically recoverable oil resources and associated production costs in the absence of a CO_2 price (non-CO_2-EOR) and in the presence of tax of $\$150/tCO_2$ (CO_2-EOR).
Source: (IEA, 2013a).

measured at standard conditions

$$R_F = \frac{N_P}{N}. \tag{2.2.1}$$

Waterflooding is the most common recovery method for new fields. Material balance and decline curve analysis are not generally useful for estimating recovery factors fields where displacement is driving production. Often the recovery factor is further broken into recovery efficiencies attributable to particular aspects of the production process. An approach used recently by BP and detailed in Smalley *et al.* (2009) encompasses both technical and economic considerations. The recovery factor is considered to consist of the product of efficiencies in pore scale displacement, E_{PS}, reservoir sweep, E_s, the connected volume factor, E_D and an economic efficiency factor E_C,

$$R_F = E_{PS} \times E_S \times E_D \times E_C. \qquad (2.2.2)$$

The application of EOR technologies is primarily to impact on either or both pore scale displacement and the reservoir sweep. Improving the connected volume factor is the domain of improved oil recovery technologies and all of these combine to impact on the economic cutoff.

2.2.2. *Limits on Microscopic Displacement Efficiency*

Subsurface permeable rocks have pore sizes of the order of 10^{-4} m or less. When fluids are moving through constrictions at this scale, interfacial forces between fluid phases and between the fluids and the solid minerals of the pore walls are important and capillarity governs the distribution of fluids at the pore scale. This is usually characterised by the dimensionless capillary number,

$$N_c = \frac{v\mu}{\sigma}, \qquad (2.2.3)$$

where v is the the linear flow velocity, μ is the displacing fluid viscosity and σ is the interfacial tension between the fluid phases. This represents the balance between interfacial and viscous forces at the pore scale and interfacial forces generally dominate in consolidated rocks when $N_c < 10^{-5}$.

In this situation non-wetting fluids will move through larger pores and through central parts of the pore space whereas wetting fluids will move along the pore walls and inhabit smaller pores. In natural rocks with mineral surfaces unaltered chemically by the presence of hydrocarbons, aqueous fluids are the wetting phase relative to oil and gas. In this case, oil moving out of the pore space of the rocks will move through the central parts of the pore space whereas resident brine can move freely along the walls of the pores. For this reason, oil ganglia can be isolated when water films "snap off" pathways to flow (Roof, 1970; Lenormand *et al.*, 1983). These isolated oil ganglia are the so called residually trapped non-wetting phase and the saturation of residually trapped oil represents the limit on what can be displaced when capillary forces dominate the microscopic fluid distribution. In water-wet rocks, this is generally in the range of 15–35% of the pore volume (Pentland *et al.*, 2010) representing between a quarter and one half of the initial oil in place (Fig. 2.2.1). The oil saturation in these systems is generally reduced to the residual saturation once a

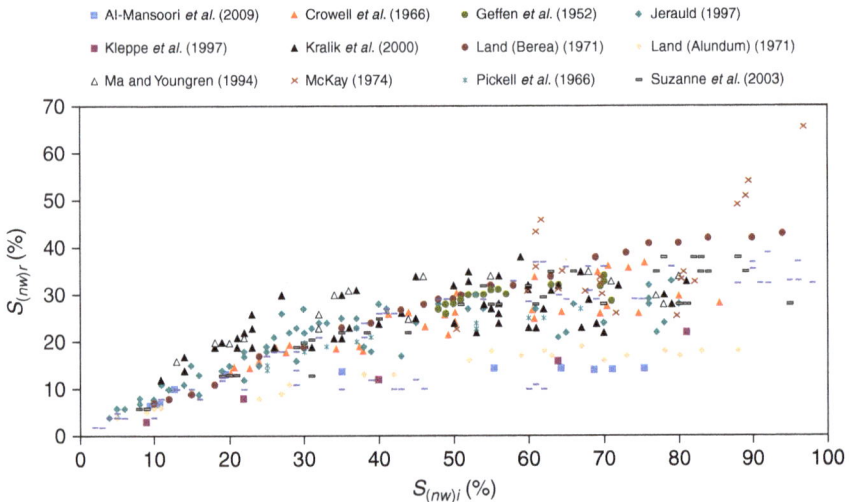

Figure 2.2.1. Compilation of initial residual trapping curves in the literature. *Source*: Pentland *et al.* (2010).

single pore volume of water has swept through the system (Salathiel, 1973; Anderson, 1987).

The presence of hydrocarbon alters the wettability of the local mineral surface to be oil wetting. Because the initially non-wetting phase will have originally inhabited the larger pores, this often leads to a scenario where the rocks are oil wet in the larger pores and water-wet in the smaller pores. This is referred to as a mixed wet system and it is thought to be the prevailing scenario in oil reservoirs. In mixed wet systems nearly all of the oil can be displaced but this requires many more pore volumes of a displacing, generally wetting, fluid to be injected than production from water-wet systems (Salathiel, 1973; Anderson, 1987).

The residual saturation is stable for a given rock so long as capillarity dominates the pore scale force balance of the fluids in the pores, i.e. $N_c < 10^{-5}$. A number of EOR techniques are understood to work primarily through the disruption of the role of capillarity in determining the local fluid distribution (Fig. 2.2.2). So-called chemical techniques use aqueous fluids for the displacement of oil but with dissolved solutes that dramatically lower the interfacial

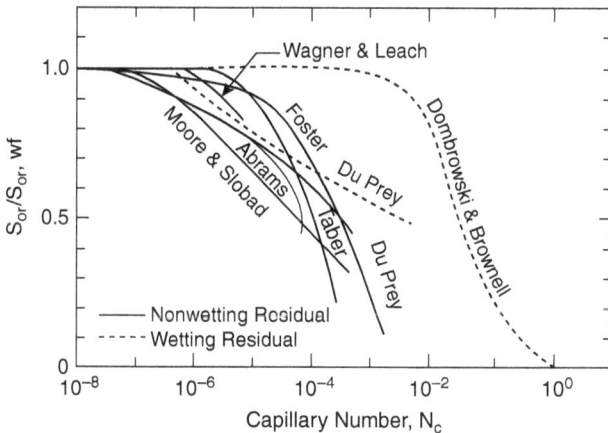

Figure 2.2.2. Residual saturation decreasing after threshold capillary numbers are exceeded.

Source: Fulcher *et al.* (1985).

tension between the water and the oil. Solvent flooding uses a displacement fluid that is miscible with the hydrocarbon phase, such as CO_2 or other hydrocarbons, thereby picking up the residual oil and recovering it with the injected solvent.

An approximate analysis can provide an estimate of how much the interfacial tension needs to be reduced to result in an effect. Typical interstitial field velocities are $v \approx 1\,\mathrm{m\,d^{-1}} \approx 10^{-5}\,\mathrm{m\,s^{-1}}$. The interfacial tension between oil and water is generally $\sigma = 6 \times 10^{-2}\,\mathrm{N\,m^{-1}}$ and the viscosity of light crude oil is similar to water, $\mu = 10^{-3}\,\mathrm{Pa \cdot s}$. This results in $N_c \approx 10^{-7}$ which suggests that increases of a factor of 100 or more are required to initiate desaturation of the residual non-wetting phase. Thus EOR techniques focused on reducing interfacial tension require chemicals that can bring interfacial tension to less than $10^{-3}\,\mathrm{N\,m^{-1}}$. It is also worth noting that a displacement at these capillary numbers also has a large impact on the other multi-phase flow properties. Relative permeability curves begin to increase and capillarity significantly weakens (Fig. 2.2.3).

The overall effect of enhancing recovery through an increase in the microscopic displacement factor is to shift the range of values $0.5 < E_{ps} < 0.75$ to $0.7 < E_{ps} < 1$.

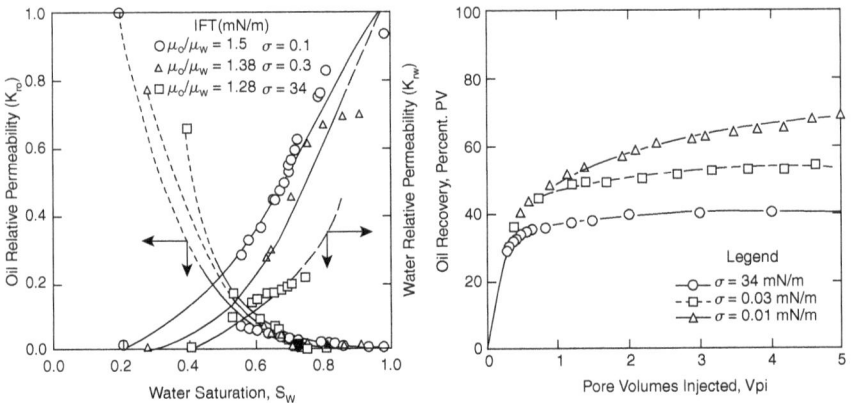

Figure 2.2.3. The impact of interfacial tension reduction on relative permeability and oil recovery in a linear core flood.

Source: Amaefule and Handy (1982).

2.2.3. *Limits on Macroscopic Displacement Efficiency*

The oil recovery factor obtained from a reservoir is always less than the microscopic displacement efficiency because the injected fluids only pass through some of the reservoir volume, even when there is good communication between the injection and production wells. This reduction in recovery is quantified by the macroscopic sweep efficiency and is a consequence of the properties of the oil and the injected fluids as well as the geology of the reservoir formation. The underlying causes are classified into three types:

(1) Heterogeneity in the fluid movement due to viscous instability;
(2) Vertical segregation of the fluids due to buoyancy effects;
(3) Heterogeneity in the reservoir rock permeability leading to channelling.

Viscous instability refers to the development of viscous fingers as shown in Fig. 2.2.4. These occur when the displacing fluid is less viscous than the oil, e.g. when gas displaces a light oil or water displaces a viscous oil. They are initiated by small scale and low level permeability fluctuations in the rock. These perturb the front between the oil and displacing fluid. The perturbations then grow because of the lower viscosity of the displacing fluid. It is important

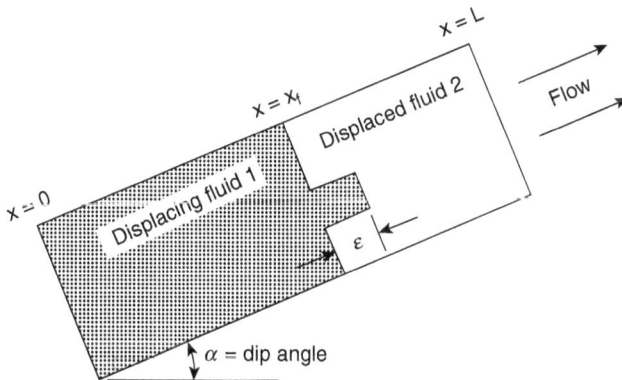

Figure 2.2.4. Conditions for viscous instabilities to form.
Source: Lake (1989).

to remember that the exact pattern of viscous fingers that will form in a given displacement is impossible to predict (they are random) although it is possible to predict their average behaviour (e.g. Todd and Longstaff, 1972; Koval, 1963; Homsy, 1987). This contrasts with channelling caused by rock heterogeneity in which the displacing fluid always follows the same path through the rock, the path being determined by the permeability pattern.

Viscous fingers are most likely to form in miscible displacements although they can also form in immiscible displacements, provided that the viscosity ratio between the displacing fluid and the oil is sufficiently high. A formal derivation of this condition can be obtained by considering small perturbations to the location of a displacement front, Fig. 2.2.4. In particular, one derives expressions for the movement of the bulk front, dx_f/dt, and compares that with the movement of the perturbation, $d(x_f + \epsilon)/dt$. The condition for instability is given by,

$$\frac{d\epsilon}{dt} = \frac{d(x_f + \epsilon)}{dt} - \frac{dx_f}{dt} > 0. \qquad (2.2.4)$$

We do not have time to cover the derivation in this module, but a remarkably simple condition falls out of this analysis. In miscible displacements, ignoring gravity, the condition for stability is dependent on the viscosity ratio

$$M = \frac{\mu_o}{\mu_d} \leq 1. \qquad (2.2.5)$$

To determine whether viscous fingers are likely to form (where μ_o is the oil viscosity and μ_d is the viscosity of the miscible displacing fluid). In immiscible displacements the condition for stability is dependent on the shock front mobility ratio,

$$M_{sf} = \frac{\frac{k_{r,w}(S_{wf})}{\mu_w} + \frac{k_{r,o}(S_{wf})}{\mu_o}}{\frac{k_{r,o}(S_{wc})}{\mu_o}} \leq 1, \qquad (2.2.6)$$

where $k_{rw}(S_{wf})$ and $k_{ro}(S_{wf})$ are the water and oil relative permeabilities at the shock front saturation, S_{wf}, and $k_{ro}(S_{wc})$ is the oil relative permeability at the connate water saturation.

Viscous fingers will form if M or $M_{sf} > 1$ and the higher the mobility ratio, the lower the sweep (Fig. 2.2.5). It is important to remember that a 1D Buckley–Leverett analysis does not capture the effect of macroscopic viscous fingering even if there is a large viscosity ratio between the oil and the displacing fluid as viscous fingering is a 3D phenomenon. It should also be noted that it is very hard to explicitly model viscous fingers using conventional reservoir simulators. You need a very fine grid and random permeability or saturation fluctuation to trigger the fingers (Fig. 2.2.6). It is more usual to capture the average effects of viscous fingers using an empirical model such as the (Todd and Longstaff, 1972) model.

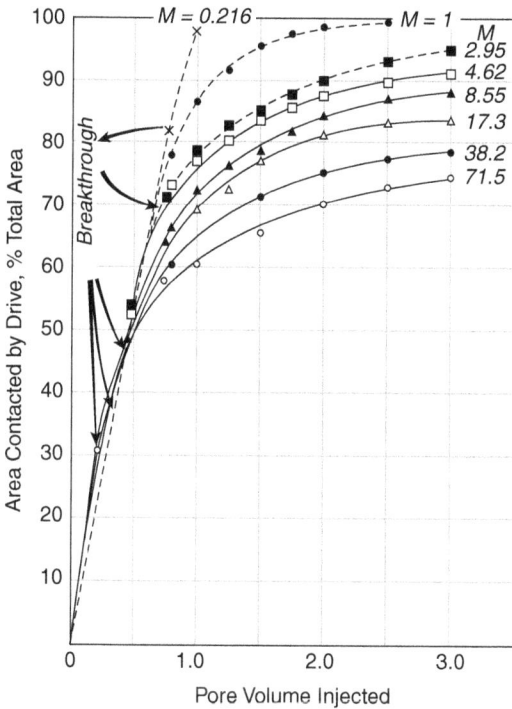

Figure 2.2.5. The impact of viscous instability on sweep for a miscible displacement in a quarter five spot pattern.

Source: Haberman (1960).

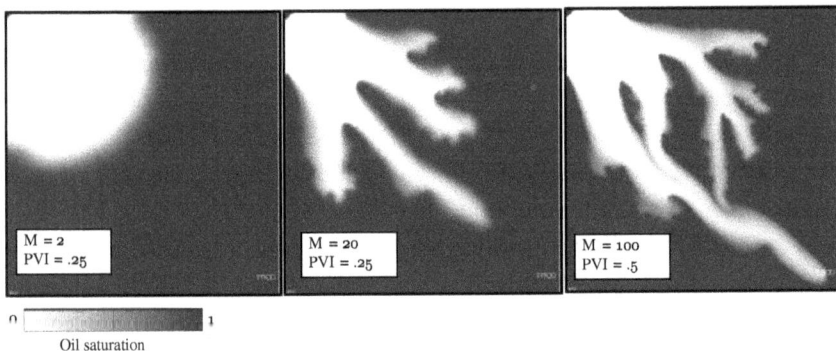

Figure 2.2.6. Viscous fingering reduces the aerial sweep in a Hele Shaw cell.
Source: Djabbarov (2014).

Gravity can enhance or stabilise viscous fingering depending upon whether it acts to increase the mobility of the displacing fluid (e.g. injecting gas underneath oil to displace the oil upwards) or to decrease its mobility (e.g. injecting gas on top of oil to displace the oil downwards). In the latter case, viscous fingers may still form unless the injection rate is sufficiently slow that gravity dominates the flow. For miscible gas injection this condition occurs when (Dumore, 1964)

$$u < \frac{\rho_o - \rho_g}{\mu_o(\ln M)} kg, \tag{2.2.7}$$

where u is the injection velocity, ρ_o is the oil density, ρ_g is the gas density, k is the permeability and g is the acceleration due to gravity. An approximate analysis also shows that velocities here are too small to be useful. Take $\Delta\rho = 400\,\mathrm{kg\,m^{-3}}$, $k = 10^{-14}\,\mathrm{m^2}$, $\mu_o = 10^{-3}$ Pa·s and $M = 10$ and this results in $u < 1.84\times10^{-3}\,\mathrm{m\,d^{-1}}$. Decreasing the mobility ratio to $M = 1.1$ only improves this to $u < 4.4 \times 10^{-2}\,\mathrm{m\,d^{-1}}$.

Thus in most cases this stable velocity results in an injection rate (and hence a production rate) that is so low that it is uneconomic, even if the reservoir structure allows the gas to be injected above the oil in this way. Instead engineers tend to focus on reducing the viscosity ratio between the injected fluid and the oil (polymer injection) or reducing the mobility of the injected fluid by, for example, injecting water and gas either simultaneously or in slugs (water alternating gas, WAG).

Gravity can also reduce macroscopic sweep if the injected fluid has a large density difference with the oil as well as a lower viscosity and is injected horizontally (Fig. 2.2.7). In this case there will be gravity segregation of the fluids (e.g. Dietz, 1953; Christie

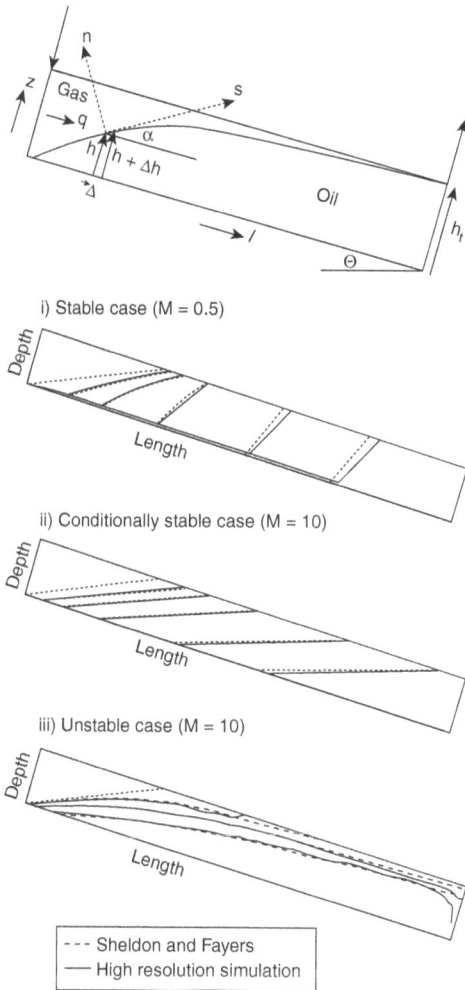

Figure 2.2.7. Conditions for gravity override in a homogeneous, but anisotropic system.

Source: From Fayers and Muggeridge (1990).

et al., 1990; Fayers and Muggeridge, 1990; Tchelepi and Orr, 1994). This is particularly important in gas injection processes and CO_2 sequestration when the injected gas migrates quickly to the top of the reservoir and then flows rapidly along the top, forming a gravity tongue (Fig. 2.2.7). This effect may be less in a dipping reservoir if the gas is injected at the top of the dipping layer and if the vertical permeability is low. It is possible to estimate whether a gravity tongue will form using the dimensionless number (Fayers and Muggeridge, 1990).

$$R_{v/g} = 2 \left[\frac{u(1 - 1/M)\mu_o}{\Delta\rho g k_z} - \theta \right] \frac{h}{L}, \qquad (2.2.8)$$

where $\Delta\rho$ is the density difference between the oil and the gas (kg m^{-3}), k_z is the vertical permeability (m^2), h is the reservoir thickness (m), L is the well spacing (m) and θ is the dip (radians). Note, this can only reliably calculate the effects of dip when $\theta < 0.1$ (about 10°). The displacement will always be stable if $R_{v/g} < 0$ (because $M < 1$ or because of a large dip). A gravity tongue will form if $0 < R_{v/g} < 1$, for $1 < R_{v/g} < 10$ flow will be influenced by both gravity and viscous effects whilst the displacement will be viscous dominated if $R_{v/g} > 10$. It should be noted that it is very difficult to use numerical simulations to model unstable gravity tongues because a high degree of vertical refinement is needed to capture the leading edge of the tongue, instead it may be better to use a vertical equilibrium option (if available in your simulator) if the above calculation suggests flow will be gravity dominated. As with viscous instability the main way to manage gravity segregation is through control of the mobility of the injected fluid, typically this is via water alternating gas injection. The water tends to reduce the mobility of the gas and, even if it does not fully prevent the formation of a gravity tongue, it will improve sweep as the water will tend to sink towards the bottom of the reservoir whilst the gas rises to the top.

Geological heterogeneity is perhaps the most usual cause of reduced macroscopic sweep although it can, on occasion, improve sweep e.g. a fining upwards sequence will reduce the tendency of gas to form a gravity tongue above oil during gas injection. EOR processes are thought to be more sensitive to geological heterogeneity because of the larger viscosity ratios between the injected fluid and the oil. This results in a higher degree to channelling through larger scale heterogeneities than would be seen in an ordinary waterflood. Smaller scale heterogeneity is believed to increase dispersion on the field scale (Mahadevan *et al.*, 2003) which can further reduce the effectiveness of EOR processes by diluting the active agent in the injected water (e.g. polymer or low salinity water) or preventing miscibility developing between injected gas and the oil. Conversely transverse dispersion can reduce the level of viscous fingering and thus improve sweep.

A major issue when attempting to quantify the impact of geological heterogeneity on macroscopic sweep is uncertainty. In most reservoirs even though geologists may have a good understanding of the depositional environment in which the reservoir rocks were laid down and there is data regarding the permeability distribution along the wellbore we simply do not know the exact permeability distribution between wells. Technically it should be possible to create a large number of possible realisations of this permeability distribution, conditioned to well data, but in practice there is often neither time or resources to be able to simulate flow through all these realisations and thus evaluate the range of possible outcomes from a chosen EOR process.

Unlike for viscous instabilities or gravity influenced flows, there are also no reliable dimensionless numbers that can be used to estimate the impact of heterogeneity on performance. Some authors advocate the use of the (Dykstra and Parsons, 1950) coefficient. This assumes that the reservoir is layered but has been extended for non-layered systems by (Jensen and Currie, 1990).

$$V_{dp} = 1 - e^{-S_k},$$

whereas,

$$S_k = \left(1 + \frac{1}{4(n-1)}\right) \sqrt{\frac{1}{n-1} \sum_{i=1}^{n} (\ln k_i - \overline{\ln k})},$$

for a log normal permeability distribution. The Lorenz coefficient (Schmalz and Rahme, 1950) is another widely used measure of heterogeneity but, to date, there is no robust heterogeneity index that both correlates with EOR performance in all types of heterogeneous reservoirs and can be calculated without resorting to a flow simulation.

The usual way to improve macroscopic sweep, if oil has been bypassed on the interwell scale and can be identified by seismic surveying, is via the drilling of infill, horizontal or multilateral wells. For smaller scale heterogeneities, flood front conformance can be temporarily improved by shutting off high permeability zones using mechanical means or polymer gels e.g. (Seright *et al.*, 2003), thereby forcing the injected fluid into lower permeability zones and thus displacing additional oil. This type of treatment is most effective if the high permeability zone is physically isolated from these lower permeability zones by continuous, impermeable zones (e.g. shales). If this is not the case then the injected fluid will simply flow around the gel and back into the higher permeability zone further into the reservoir (Sorbie and Seright, 1992). Mitigating the adverse effects of geological heterogeneity on EOR and, in particular, reducing the uncertainty in what these effects will be remains one of the most significant challenges for reservoir engineers.

In summary, macroscopic sweep can be improved through mobility control, either of injected water into particularly viscous oil or in the control of gas being injected for miscible displacement floods. In the latter case, both viscous instability and gravity override can be controlled through the application of WAG. The promise of EOR in this area is to improve values of macroscopic sweep from $E_S = 0.6$ to upwards of $E_S = 0.8$. Combining enhancements in pore scale displacement and macroscopic sweep efficiencies improves recovery factors from at least $R_F = 0.3$–0.8. This is promising on the one hand. On the other, this does not consider the impacts of economic cutoffs and compartmentalisation. Thus recoveries achieving $R_F > 0.6$, such as the Prudhoe Bay field off the N. Slope of Alaska are impressive indeed.

2.3.   Gas Injection

While the injection of water to maintain pressure in a reservoir and displace oil is an effective recovery mechanism, as discussed in Sec. 2.2, capillary trapping results in maximum microscopic displacement efficiencies of only 50–75% of the oil initially in place. The injection of fluids that are miscible with the oil phase, however, means that the microscopic displacement efficiency can be reduced to a complete recovery of the oil. This is the basis for solvent gas injection EOR, where the term "gas" here is used loosely to refer to a light non-polar fluid phase such as CH_4 or CO_2 that are gases at atmospheric conditions. At the depths at which gas injection is generally applied, these fluids, and CO_2 in particular, are not in the gaseous state.

Quantitative descriptions of chemical component transport, fluid phase displacement and fluid phase chemical composition are the aims of a study of gas injection processes. The displacement processes can involve up to three fluid phases and require representation of three or four constituent components that can transfer between the phases. This makes their description highly complex and generally within the domain of numerical simulation. Detailed references on gas injection processes include (Lake, 1989) and (Orr, 2007) and the latter provides a comprehensive set of analytical solutions to two-phase multi-component displacements. In this lecture, we will only learn some of the basic concepts underlying the physics of gas injection. I will first discuss qualitatively the representation of miscibility on ternary phase diagrams. A brief refresher of the Buckley–Leverett formulation of the flow equations for two-phase immiscible displacement will be used as a starting point to develop the equations for a two-phase multi-component system with miscibility. These tools will then be applied in one of the simpler examples of a miscible displacement process, a 1D first contact miscible water alternating gas injection problem.

2.3.1. *Phase Equilibrium for Gas Drives*

When a gas like CO_2 or CH_4 is injected into an oil phase a number of chemical component transfers will take place between the fluids. Parts of the oil will vapourise into the gas-rich phase and parts of the gas phase will dissolve into the oil. This changes the chemical composition of each fluid phase and as each phase contacts "fresh" oil or gas a new set of chemical transfers will take place changing the composition even further. These stages of phase contact, chemical equilibrium, movement, contact and reequilibrium have major impacts on the flow of each through the controlling relationship between the chemical composition of a phase and its saturation.

The relationship between phase saturation and chemical composition for oil–gas systems is conventionally illustrated on a ternary diagram. We will approach this in two parts — first, we will discuss the use of a ternary diagram to represent the overall composition of a system in terms of three chemical components. Then, we will separately introduce the relationship between chemical component composition and phase equilibrium in simpler, two component systems. Finally, we will bring the two concepts together in a three-component two-phase ternary diagram.

Figure 2.3.1 shows a ternary diagram for a single-phase system. The vertices of the diagram are the three chemical components. Conventionally a lighter component, C_1 is the top vertice, an intermediary component, C_2 is on the right and a heavy component C_3 is on the left. For an example with a real system, see Fig. 2.3.8 and references (Orr *et al.*, 1981; Gardner *et al.*, 1981). Often the intermediate and heavy components are *pseudo components*, an arbitrary division of the crude oil into light and heavy fractions. A fluid phase made up of various mole fractions of each component is represented by a point within the triangle. The lines in the triangle represent lines of constant mole fraction for one of the components.

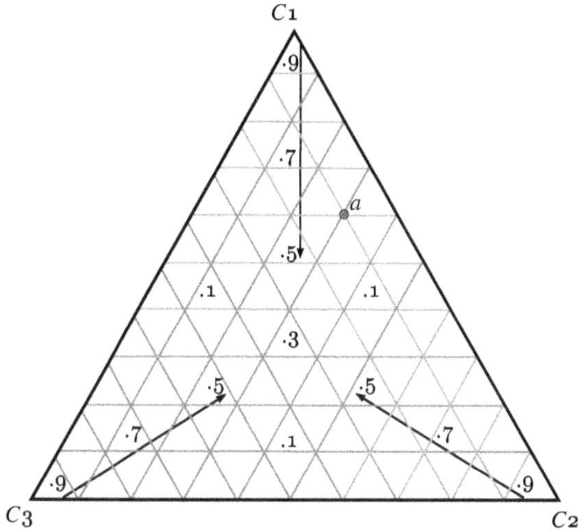

Figure 2.3.1. A single-phase ternary diagram. A fluid with composition a is composed of 60% C_1, 30% C_2, and 10% C_3.

Thus, a fluid with composition a is made of 60% C_1, 10% C_3 and 30% C_2. Can you spend a few minutes right now and mark in the notes the following compositions:

- b — 20% C_1, 50% C_2 and 30% C_3;
- c — 50% C_1, 20% C_2 and 30% C_3;
- d — 30% C_1, 10% C_2 and 60% C_3.

When multiple phases exist there will be a region of the diagram in which the overall chemical composition can only be represented by the sum of the chemical compositions of two coexisting but separate phases.

Consider first, however, a pressure-composition diagram for a two-component system, CO_2 and n-hexane at 333.15 K, shown in Fig. 2.3.2. In this system, there are potentially two phases, a liquid hexane rich phase and a gaseous CO_2-rich phase, and there are two components, hexane and CO_2. If we refer to CO_2 as component one, C_1, and hexane as component two, C_2, we can see that $C_1 = 1 - C_2$. Similarly, if we refer to the liquid phase as phase one and the gaseous phase as phase two, the respective saturations, S_i are related

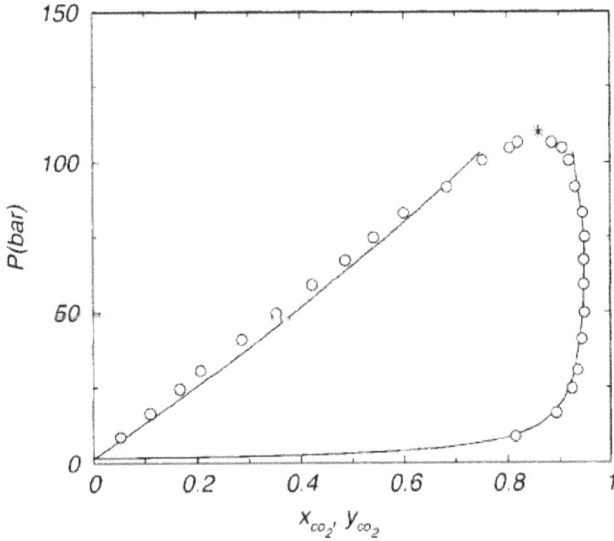

Figure 2.3.2. Vapour–liquid equilibria for mixtures of CO_2 with n-hexane at 333.15.

Source: Potoff and Siepmann (2001).

through $S_1 = 1 - S_2$. You can see on the diagram that to the left of the bubble point curve, $S_1 = 1$ and to the right of the dew point curve $S_2 = 1$. Within the envelope, the two phases are in equilibrium although the specific mole fractions of each phase are dependent on the overall chemical composition of the system. Consider a fluid mixture with overall chemical composition within the two-phase envelope. The chemical composition of the liquid part of the mixture is given by the bubble point chemical composition, c_{11}, the concentration of chemical component one in phase one. The chemical composition of the vapour part of the mixture is given by the dew point chemical composition, c_{12}, the concentration of the chemical component one in phase two. The constraints of the specific liquid and vapour chemical compositions along with the total overall concentration of the chemical components results in a determination of the saturation of liquid and vapour. In other words, the total concentration of component one is given by,

$$C_1 = c_{11}S_1 + c_{12}S_2 = c_{11}S_1 + c_{12}(1 - S_1). \tag{2.3.1}$$

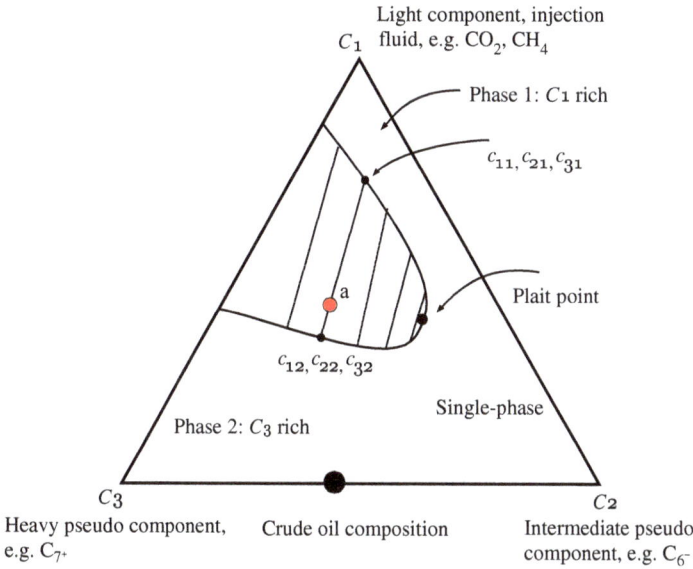

Figure 2.3.3. A two-phase ternary diagram. A fluid mixture with overall composition a will exist as a mixture of two phases with component concentrations given by the ends of the tie line on the binodal curve.

Thus the saturation is determined by,

$$S_1 = \frac{C_1 - c_{12}}{c_{11} - c_{12}}. \qquad (2.3.2)$$

A similar analysis applies to a three component system with two phases represented in a ternary diagram. In the ternary diagram in Fig. 2.3.3, a phase rich in the light component is fully miscible with a phase rich in the heavy component and the same is true between the intermediate and heavy components. There is a region of immiscibility, however, between phases rich in the light and intermediate components. An overall fluid mixture composition in this two-phase region given by the point a would be made of a corresponding amount of C_1, C_2 and C_3. No single-phase with this particular composition is stable on this diagram and thus C_1 would represent the sums of the concentration of component 1 in phase 1, c_{11}, multiplied by the saturation of phase one, S_1, and the concentration of the component in phase two, c_{12}, multiplied by the saturation of phase two, S_2,

$$C_1 = c_{11}S_1 + c_{12}S_2 = c_{11}S_1 + c_{12}(1 - S_1). \qquad (2.3.3)$$

As with the binary component system, there is a line going through the overall chemical component composition and intercepting the phase boundaries at the locations where the chemical composition of the phases is in equilibrium with the the total chemical composition. In this case, however, it is not just a function of pressure, but also the total chemical composition. In any case, for any given ternary chemical composition, C_1, C_2 and C_3 within the two-phase envelope, there are specific phase chemical compositions, $c_{11}, c_{21}, c_{31}, c_{12}$, etc. associated with that overall composition. The line connecting the phase boundaries with the chemical composition within the two-phase envelope is known as a tie line.

Similar equations could be written for components two and three. Note that all ternary diagrams only apply at a single pressure and temperature, and the compositions of the fluids at the end of the tie lines, e.g. c_{11}, c_{12} for component C_1, are a function of the total mixture concentration only, C_1, C_2, C_3 at this fixed pressure and temperature. This is determined by the rules of thermodynamic equilibrium of the two phases and thus the chemical composition of each phase, e.g. c_{11} and c_{12}, can be thought of as constants determined by the total chemical composition at the given pressure and temperature.

An important result of this is that the respective saturations of the two phases is determined by the total chemical composition. This is seen by rearranging Eq. (2.3.3),

$$S_1 = \frac{C_1 - c_{12}}{c_{11} - c_{12}}. \qquad (2.3.4)$$

The saturation could be equivalently found through similar equations in terms of C_2 or C_3. In other words, on a given tie line, the saturation of the fluids is entirely determined by one total chemical component concentration.

Only a few potential tie lines are shown in Fig. 2.3.3, but there will be a tie line associated with any composition inside the two-phase envelope with corresponding phase compositions and saturations.

The other significant point highlighted on the figure is the *Plait point*. This is the point at which the length of the tie lines vanish to zero. One way to think about this is in terms of "developing miscibility". A phase initially rich in C_1 such that it coexisted with a phase rich in C_2 could be diluted with contributions of the component C_3 until the total composition moved out of the two-phase region and the fluids were miscible. This turns out to be a key process during gas injection that we will now explore further.

Consider first a gas injection process represented by the ternary diagram shown in Fig. 2.3.4. The resident oil has a composition given by the point on the lower right of the diagram and the injection gas is entirely composed of component C_1. This gas is fully miscible with the resident oil because the mixing line between the two end member chemical compositions (injected gas and resident oil) does not pass through the two-phase region. This process is called *first contact miscible* because the injected gas is already miscible with the resident oil upon first contact between the two fluids.

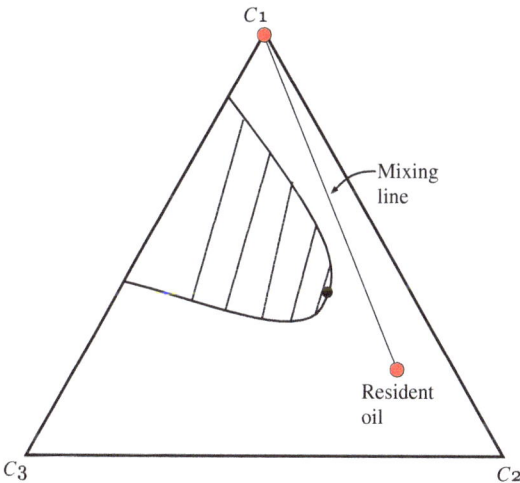

Figure 2.3.4. A two-phase ternary diagram showing the mixing line for the injection of C_1, first contact miscible with the resident oil.

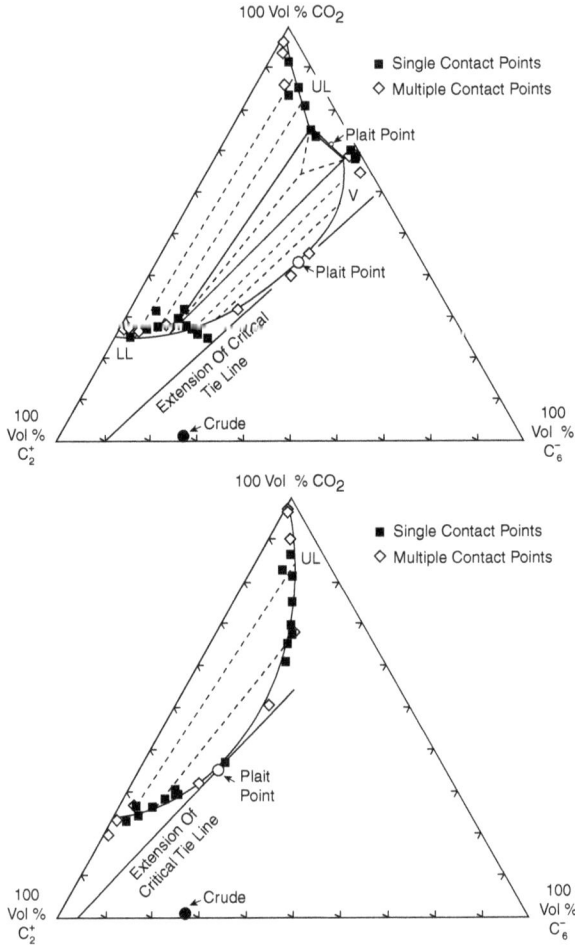

Figure 2.3.5. Ternary diagrams of mixtures of CO_2 with Wasson crude at 1,350 psi (above) and 2,000 psi (below). Both are at 105°F. Increasing pressure results in a larger field of potential miscibility with the injected fluid.

Figure 2.3.5 shows a ternary diagram with two-phase regions for a mixture of CO_2 with Wasson crude oil at two pressures and 105°F.

In the case that the injected fluid is not initially fully miscible with the resident oil, there are dilution pathways along which the fluids can develop miscibility. These are called multi-contact miscible processes for reasons that will become apparent shortly. Consider

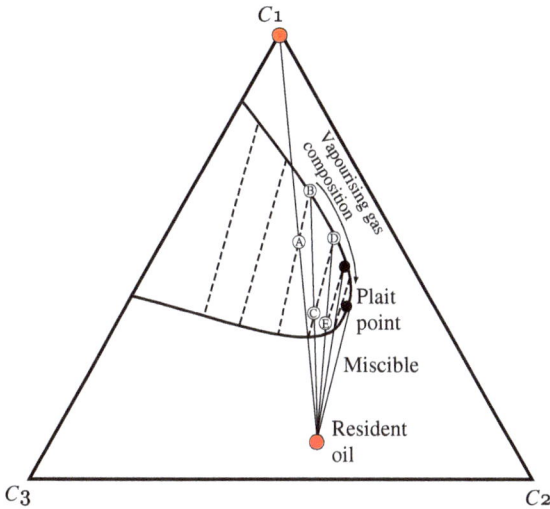

Figure 2.3.6. A two-phase ternary diagram showing the development of miscibility through a vapourising gas drive, injecting C_1, multi-contact miscible with the resident oil.

first a system where miscibility is developed as the injected gas incorporates increasing amounts of heavier elements of the resident oil into the gas-rich phase. This is shown in Fig. 2.3.6. Injecting a gas made entirely of component C_1 into the resident oil results in a mixture at point A that falls within the two-phase region. The mixture quickly equilibrates to form a gas-rich phase with a new composition given by point B and an oil-rich phase on the opposing end of the tie line. The new gas-rich phase moves further to contact more fresh resident oil and forms a mixture with overall chemical composition given by point C on the diagram. This again falls in the two-phase region resulting in the formation of a gas-rich phase now with the composition given by point D. You can see from the diagram that as the gas-rich phase continues to contact more of the uncontacted resident oil, it continues to pick up amounts of C_2 and C_3 such that the gas phase composition evolves until miscibility has been achieved. Because of the multiple contacts between newly equilibrated gas phase and unconnected oil, this is known as multi-contact miscibility. In addition to this being a multi-contact miscible

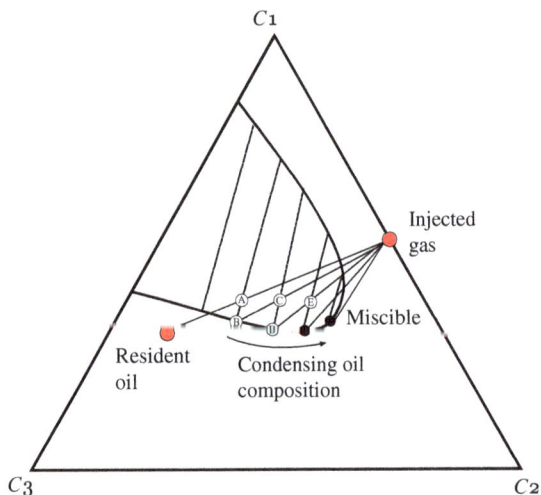

Figure 2.3.7. A two-phase ternary diagram showing the development of miscibility through a condensing gas drive, injecting a fluid with components C_1 and C_2, multi-contact miscible with the resident oil.

process, this particular example is known as a *vapourising gas drive* because the mechanism for the development of miscibility is to increasingly vapourise components of the oil into the gas-rich phase.

There is another mode of multi-contact miscibility known as the *condensing gas drive*. Consider the gas injection process shown in Fig. 2.3.7 where a gas, now containing fractions of components C_1 and C_2, is injected into the resident oil with a composition on the lower left side of the ternary diagram. The first contact produces a mixture in the two-phase region given by the overall composition at point A. This equilibrates to two phases with chemical compositions given by the ends of the tie lines. In this scenario, the gas-rich phase is moving away from miscibility but the oil-rich phase, condensing components from the gas, is moving towards miscibility as it vapourises the oil components. In other words, the equilibrated oil with chemical composition at point B is closer to miscibility with the injected gas. As more gas is injected, the oil with this composition will again mix forming a mixture with composition at point C. This mixture will separate into two phases and the oil rich component will have a chemical composition given by point D, closer to miscibility with the injected gas. Eventually, the oil-rich phase will condense enough components out of the gas such that it has miscibility with subsequently injected gas.

To generalise this discussion, consider the potential gas injection scenarios possible with gas compositions g_1 or g_2 and reservoir oil compositions r_1 or r_2 shown in Fig. 2.3.10. Take a few minutes to answer the following questions using the diagram.

Classify the following injection scenarios as first contact miscible, a vapourising gas drive, a condensing gas drive, or immiscible. If this is a a multi-contact miscible process, show some of the steps to the development of miscibility.

- The injection of gas with composition g_2 into the oil with composition r_2;
- The injection of a gas with composition g_1 into the oil with composition r_2;
- The injection of a gas with composition g_2 into the oil with composition r_1;
- The injection of a gas with composition g_1 into the oil with composition r_1.

The injection of gas with composition g_2 into the oil with composition r_2 would be a first contact miscible process because the mixing line between the fluids does not pass through the two-phase region. The displacement of reservoir fluid r_2 with gas of composition g_1 would be multi-contact miscible and result in a vapourising gas drive as the gas-rich phase approaches miscibility with the resident oil. Similarly the injection of gas with composition g_2 into oil with composition r_1 would be multi-contact miscible and result in a condensing gas drive as the oil-rich phase approaches miscibility with the injected gas. If you considered the injection of gas g_1 into reservoir oil with composition r_1 and followed the same procedure as for the previous multi-contact miscible systems you would find that miscibility would not be approached for this system. This would thus be classified as an immiscible flood. In order for miscibility to develop, the initial compositions of the gas and oil phases need to be on opposite sides of the *critical tie line*. This is an imaginary tie line that goes through the Plait point but is parallel to the tie lines in the two-phase region just before the Plait point is reached.

Through this simple graphical analysis of a gas injection process it is thus possible to tell a lot about the system, whether there are opportunities for a vapourising or condensing gas drive and if the composition of an injected gas could be tailored, by diluting for example with some amount of the component C_2, so that miscibility develops. At the same time it is important to keep in mind the limitations of such a simplistic analysis. Ternary diagrams only apply at a single pressure and temperature and we have not discussed at all the significant efforts required through modelling and in the laboratory to obtain such diagrams, particularly where crude oil is concerned. The multiple contact process is not a discrete process as was shown here, rather the gas-rich or oil-rich phases are continuously evolving as they contact more resident oil or injected gas. Fig. 2.3.9 shows the composition path for a displacement of crude oil by CO_2 simulated in a 1D displacement scenario (Gardner *et al.*, 1981). Our analysis was 0D and did not consider how fluids flow and contact each other, nor any time constraints on the processes of chemical

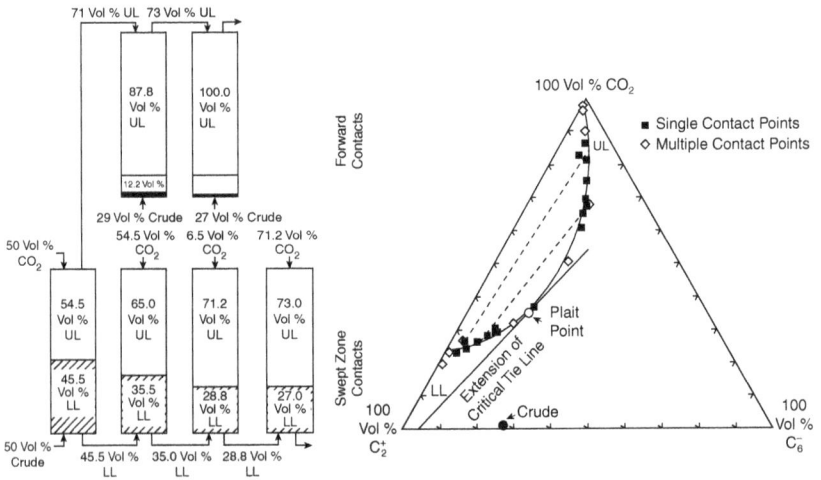

Figure 2.3.8. A multi-contact mixing experiment performed with Wasson crude oil and CO_2 at 2,000 psi and 105°F.

Source: Gardner *et al.* (1981).

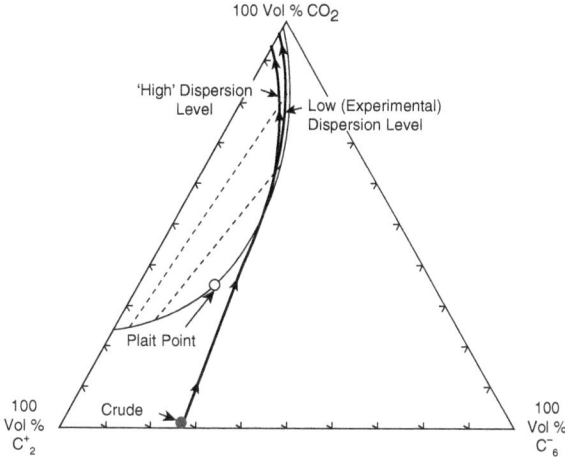

Figure 2.3.9. A two-phase ternary diagram from Gardner *et al.* (1981) showing the simulated composition path in a 1D displacement for the process shown in Fig. 2.3.8.

equilibrium. These were thus flash calculations in which we assumed that equilibrium was instantly reached and that evolving gas rich fluids were immediately plucked from the mixtures and moved on to contact new oil once the equilibrium has been (instantly) achieved. There is some basis for this as the gas rich fluids have higher mobility than the oil rich fluids. Thus, what is actually observed in a laboratory core flood or field scale process is described only qualitatively by this analysis. In practice, for example, the mixing line for the injection of CO_2 into a hydrocarbon needs to be well above the Plait point in order for miscibility to develop in a reservoir setting. Perhaps most important, this analysis tells us nothing about how the fluids themselves will flow in such a process. How much oil will be recovered and how quickly? How much solvent will be required to recover a particular amount of oil? Some of these questions will now be addressed in the next section.

2.3.2. *Transport in Two-Phase Multi-Component Systems*

I provide a brief review here of the Buckley–Leverett solution to immiscible displacement, which we will then extend to incorporate

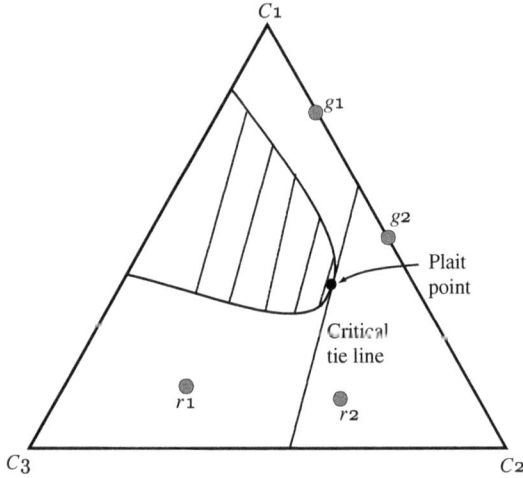

Figure 2.3.10. A two-phase ternary diagram to be used for exercises.

multi-component fluids with miscibility (Fig. 2.3.10). The following presentation has been derived primarily from (Pope, 1980; Lake, 1989; Orr, 2007; Blunt, 2013).

2.3.2.1. *Review of the Buckley–Leverett solution*

The Buckley–Leverett equations provide a solution to a problem of the displacement of one fluid by another immiscible fluid in one dimension. In this system, there is no chemical mass transfer between the phases and thus each fluid phase can be thought of to exist entirely as a single component. Using fractional flow theory the governing mass balance expressed in dimensionless form is given by

$$\frac{\partial S_i}{\partial t_D} + \frac{\partial f_i}{\partial x_D} = 0, \tag{2.3.5}$$

where S_i is the saturation of a phase i. The fractional flow, f_i, of phase i is the ratio of the flow rate of that phase, q_i, to the sum total of all fluid flow rates, $q_t = \sum_i q_i$,

$$f_i = \frac{q_i}{q_t}.$$

Dimensionless time t_D and distance x_D can be defined by,

$$t_D = \frac{q_t t}{\phi L},$$

$$x_D = \frac{x}{L},$$

where L is a reference length scale and ϕ is the porosity of the system.

In fractional flow theory, capillary pressure gradients are usually assumed to contribute little to the flow. If we also ignore gravity (i.e. consider flow in the horizontal direction) f_i can be expressed as,

$$f_i = \frac{\lambda_i}{\sum_i \lambda_i},$$

where the mobility of a phase, λ_i is the ratio of the relative permeability of that phase at the phase saturation, $k_{r,i}(S_i)$ to the viscosity of the phase, μ_i,

$$\lambda_i = \frac{k_r(S_i)}{\mu_i}.$$

Thus the fractional flow is a function of saturation and Eq. (2.3.5) can be written as

$$\frac{\partial S_i}{\partial t_D} + \frac{df_i}{dS_i}\frac{\partial S_i}{\partial x_D} = 0. \qquad (2.3.6)$$

This type of equation lends itself to solution by the method of characteristics. Using this technique, solutions of the form $S_i(x_D, t_D)$ are not obtained but rather characteristic curves along which S_i is a constant value, e.g. $x_D = f(t_D)|_{S_i=S_o}$. The solution is

$$x_D = \frac{df_i}{dS_i}(S_i)t_D, \qquad (2.3.7)$$

which says that a saturation, S_i, will propagate forward in the medium at a rate given by the derivative of the fractional flow function, df_i/dS_i, evaluated at that saturation. The derivative of the fractional flow function is thus referred to as the dimensionless velocity of the saturation,

$$v_D = \frac{df_i}{dS_i}.$$

The derivative of the fractional flow curve, however, is often a non-monotonic function in which case the saturation associated with a given propagation velocity is non-unique. From a physical standpoint this is impossible and results in the formation of a shock. In this case, Eq. (2.3.7) provides the saturation velocities for all of the saturations greater than the shock saturation, which I will denote as S^*. Saturations below S^* have velocities equal to the shock saturation. The shock saturation and velocity are obtained through mass balance considerations and given by the equation

$$v_D^* = \left.\frac{df}{dS_1}\right|_{S_1=S^*} = \left.\frac{f_1 - f_1^o}{S_1 - S_1^o}\right|_{S_1=S^*}, \tag{2.3.8}$$

where S_1^o and f_1^o are the saturation and corresponding fractional flow of phase 1 at an initial condition prior to displacement. In a graphical representation, S^* is the saturation at which the slope of the fractional flow curve is equal to the slope of the straight line passing through the points (S_1^o, f_1^o) and $[S_1^*, f_1(S_1^*)]$.

2.3.2.2. *Two-component, two-phase displacement*

Consider now a system with multiple phases and multiple chemical components that can move between the phases such that the component concentration of a chemical i in phase j can be given by c_{ij}. In Sec. 2.3.1, we considered systems of two phases, $j = 1, 2$, and three chemical components $i = 1, 2, 3$.

The total concentration of the component, C_i, will be given by the sum of the product of the fraction of that component in each phase with the saturation of those phases,

$$C_i = \sum_j c_{ij} S_j. \tag{2.3.9}$$

Similarly, the total fractional flux of a chemical component i is given by the sum of the products of the concentrations in each phase with the fractional flow of that phase,

$$F_i = \sum_j c_{ij} f_j. \tag{2.3.10}$$

The rate of change of mass of a chemical component per unit volume is given by the sum of the rates of change in each of the phases,

$$\phi \frac{\partial}{\partial t} \sum_j c_{i,j} S_j \tag{2.3.11}$$

$$= \phi \frac{\partial C_i}{\partial t}. \tag{2.3.12}$$

This must be balanced by fluxes of the component into and out of the volume. In this class we will consider a 1D system where the net flux is given by,

$$\frac{\partial}{\partial x} \sum_j c_{i,j} q_j \tag{2.3.13}$$

$$= \frac{\partial}{\partial x} \sum_j c_{i,j} f_j q_t \tag{2.3.14}$$

$$= q_t \frac{\partial F_i}{\partial x}, \tag{2.3.15}$$

for incompressible fluid systems where $\frac{\partial q_t}{\partial x} = 0$. Note that volume is often not conserved when components transfer phases, i.e. the density of CO_2 in a hydrocarbon rich phase may well be very different than the density of CO_2 in the CO_2 rich phase, and so this is not always appropriate.

The conservation equation is thus,

$$\phi \frac{\partial C_i}{\partial t} + q_t \frac{\partial F_i}{\partial x} = 0. \tag{2.3.16}$$

We can convert this to dimensionless units using the chain rule,

$$\frac{\partial C_i}{\partial t} = \frac{\partial C_i}{\partial t_D} \frac{\partial t_D}{\partial t} = \frac{q_t}{\phi L} \frac{\partial C_i}{\partial t_D}, \tag{2.3.17}$$

$$\frac{\partial F_i}{\partial x} = \frac{\partial F_i}{\partial x_D} \frac{\partial x_D}{\partial x} = \frac{1}{L} \frac{\partial F_i}{\partial x_D}. \tag{2.3.18}$$

The dimensionless mass balance for the chemical component in 1D is given by

$$\frac{\partial C_i}{\partial t_D} + \frac{\partial F_i}{\partial x_d} = 0. \qquad (2.3.19)$$

Consider now a two-phase, two-component system, $i = 1, 2$ and $j = 1, 2$. Assuming a two component system is like assuming that our composition path is constrained to a single tie-line. In a two-component system, as on a specific tie line, the saturation of a phase, S_j, and the chemical component concentration in that phase, c_{ij}, in this system are functions of the total chemical concentration, C_i only. The phase concentrations are given by the ends of the tie line associated with C_i and the saturation is determined by the location of C_i on the tie line, e.g. Eq. (2.3.4). As a result,

$$F_i = \sum_j c_{ij} f_j = \sum_j c_{ij}(C_i) f_j(S_j) = F_i(C_i), \qquad (2.3.20)$$

and Eq. (2.3.19) can be written as

$$\frac{\partial C_i}{\partial t_D} + \frac{dF_i}{dC_i} \frac{\partial C_i}{\partial x_d} = 0. \qquad (2.3.21)$$

This is the same type of equation as Eq. (2.3.6), describing the flow of the fluid phases in the Buckley–Leverett formulation, and thus can be solved using the method of characteristics for the velocity at which a total concentration wave propagates in a two-phase system,

$$x_D = \frac{dF_i}{dC_i} t_D. \qquad (2.3.22)$$

The dimensionless component concentration velocity and the shock velocities are again given as the derivative of the fraction flow function,

$$v_D = \frac{dF_i}{dC_i}, \qquad (2.3.23)$$

$$v_D^* = \frac{F_i - F_i^o}{C_i - C_i^o}. \qquad (2.3.24)$$

Component fractional flow curves (plots of F_i versus C_i) can be constructed using data describing the phase saturation as a function of chemical composition, e.g. Eq. (2.3.4), and the relative permeability curves of the two phases. We will not take this any further here, however. We will use some of this analysis in a simple example analysing the properties of first contact miscible water alternating gas displacement below. The key idea to see is that total chemical component concentration waves have velocities determined by a component fractional flow curve, $F_i(C_i)$, which are related to but distinct from fluid phase saturation wave velocities that are determined by the fluid phase fractional flow curve, $f_j(S_j)$. This concept is general to systems of multiple components whether those components are in the oil phase (as here) or in the water phase (as in chemical flooding).

2.3.3. *Fractional Flow Theory and Water Alternating Gas Injection*

One of the simplest examples of a multi-component and multi-phase displacement process is that of a first contact miscible WAG injection, where a solvent gas, first contact miscible with oil in the reservoir, is coinjected or injected in alternating slugs with an aqueous phase, into a reservoir that contains a mixture of an aqueous phase and oil. This is done with the aim of capturing the benefits of miscible gas injection, increasing the microscopic displacement efficiency to 100%, while mitigating the impact of the generally high mobility of solvent gases on fingering and gravity segregation through a reduction in the relative permeability.

Consider a 1D displacement wherein water and a hydrocarbon phase bearing a solvent (the "gas") are injected at a fractional flow given by f_w^o into a reservoir with an initial water and oil saturation at the connate water saturation given by S_w^o. There are only two phases in this system, a hydrocarbon bearing phase and the aqueous phase which I will denote with subscripts o and w respectively. Additionally, there is no chemical component exchange between the two phases and thus the saturation of the phases is not controlled directly by the

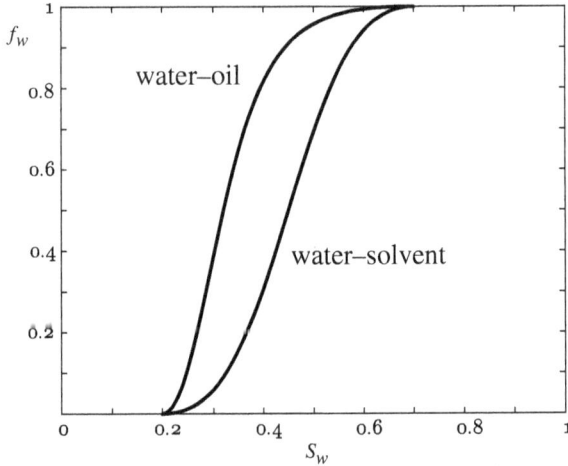

Figure 2.3.11. Fractional flow curves for an oil phase displacing water where the water–solvent curve has a higher mobility ratio.

overall chemical composition of the mixture. The speed at which the saturation waves propagate is determined by the relevant fractional flow curves. The injected oil phase with solvent will have a lower viscosity and there are two fractional flow curves for determining saturation wave velocities, the curve corresponding to the original oil–water system and the curve corresponding to the solvent oil–water system. These are shown in Fig. 2.3.11.

Thus in this system there are two phases plus two components in the oil phase, the solvent chemical C_1 and a pseudo component consisting of the remainder of the oil. The aqueous phase will be assumed to consist of a single component that has no solubility in the oil phase. Thus the total concentration of the solvent, C_1, is given by,

$$C_1 = c_{1w}S_w + c_{1o}S_o = c_{1o}S_o = c_{1o}(1 - S_w). \tag{2.3.25}$$

Similarly, the total fractional flux of the solvent, F_1, is given by

$$F_1 = c_{1w}f_w + c_{1o}f_o = c_{1o}(1 - f_w). \tag{2.3.26}$$

As a result, the mass balance Eq. (2.3.19) becomes

$$\frac{\partial[c_{1o}(1 - S_w)]}{\partial t_D} + \frac{\partial[c_{1o}(1 - f_w)]}{\partial x_d} = 0$$

$$= (1 - S_w)\frac{\partial c_{1o}}{\partial t_D} - c_{1o}\frac{\partial S_w}{\partial t_D} + (1 - f_w)\frac{\partial c_{1o}}{\partial x_D} - c_{1o}\frac{\partial f_w}{\partial x_D}$$

$$= \frac{\partial c_{1o}}{\partial t_D} + \frac{1 - f_w}{1 - S_w}\frac{\partial c_{1o}}{\partial x_D} = 0.$$

The dimensionless velocity of the solvent, $v_{D,s}$, is given by,

$$v_{D,s} = \frac{1 - f_w}{1 - S_w} = \frac{f_o}{S_o}. \tag{2.3.27}$$

This is seen to be correct when considering that the solvent does not undergo any mass exchange with the water and tracking the solvent is akin to tracking a chemical component in a single phase. In this circumstance the dimensionless velocity is simply given by the fractional flow divided by the saturation.

The water front velocity is given by the same considerations as with the Buckley–Leverett formulation, except in this case, $f_w \neq 1$. If the fractional flow is such that there is no shock and rarefaction, the velocity is given by,

$$v_{D,w} = \frac{f_w}{S_w - S_w^o}. \tag{2.3.28}$$

In the following discussion we will only consider situations where saturation changes occur entirely within a shock front.

These equations lend themselves to a graphical characterisation of the system on plots of the fractional flow versus saturation for the various fluid pairs. In the following discussion we will use such graphs to determine the velocities of (1) water shock fronts where the saturation changes and (2) solvent fronts where the system changes from an oil–water displacement system to a oil/gas–water displacement system. Sometimes water shock fronts will occur at locations distinct from the arrival of the injected gas and many times a shock front will also be associated with the arrival of the injected gas.

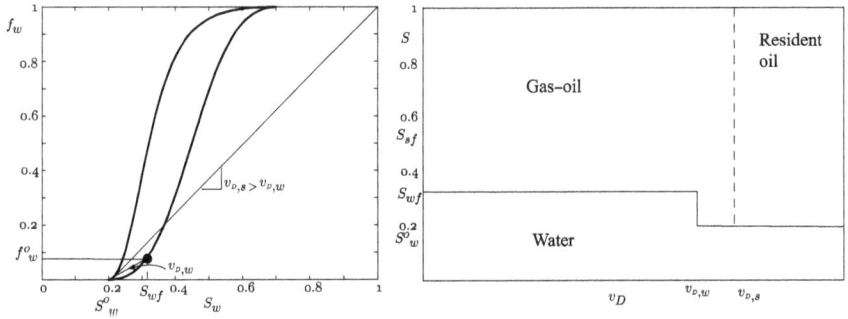

Figure 2.3.12. Showing the solvent and water injection velocities at an injection fractional flow where the solvent velocity will be larger than the injected water velocity

Consider a fractional flow f_w^o shown in Fig. 2.3.12. The water saturation associated with this fractional flow is shown as a solid point in the graph. In this case, very little water is injected relative to the solvent mixture. The solvent velocity, Eq. (2.3.27), is normally given by the slope of the line from the top right of the graph to the saturation point corresponding to the injection fractional flow. The slope of such a line would be steeper than the slope of the corresponding line from the point $(S_w^o, 0)$ to $(s_{w,f}, f_w^o)$, or alternatively, the slope of the line from the initial saturation to the location on the oil–water fractional flow curve, either of which would indicate the velocity of the associated water shock front. This suggests that the injected gas moves faster than the water shock front. Moreover, in this particular case, the gas–oil system is moving into a resident oil system that is already at the maximum water saturation. Thus, the arrival of the gas–oil front will not result in a change in water saturation which is already at the minimum (recall that the initial condition of the reservoir was at the connate water saturation). Additionally, the velocity of this front is not given by Eq. (2.3.27) but rather by the value $1/S_w^o$. Eventually, a change in water saturation occurs corresponding to the equilibrium associated with the fractional flow. The velocity of this is given by Eq. (2.3.28), the slope of the line from the point on the fractional flow curve at

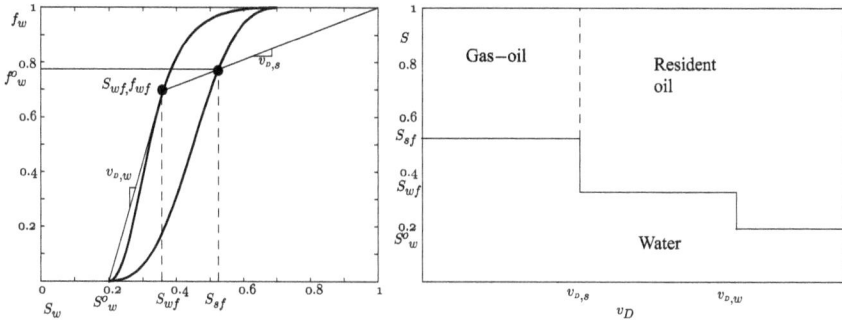

Figure 2.3.13. Showing the solvent and water injection velocities at an injection fractional flow where the solvent velocity will be less than the injected water velocity.

the initial water saturation (in this case $f_w = 0$ at this point) to the saturation point on the solvent–water fractional flow curve. The velocity of the solvent in this region is given by Eq. (2.3.27), but this is of less importance.

Now consider the system with an injection ratio with a high fraction of water, f_w^o shown in Fig. 2.3.13. The gas velocity is given by Eq. (2.3.27) but it is apparent that the water saturation wave will move faster than the solvent. Thus, initially, the water saturation will increase even though the gas has not yet arrived. An increase in water saturation in the absence of the arrival of the gas indicates that this wave velocity will be determined by the oil–water fractional flow curve. Thus, there will be some saturation, S_{wf}, on the oil–water fractional flow curve (S_{wf}, f_{wf}) associated with the wave that arrives prior to the injected gas. This saturation will not necessarily be equivalent, however, to the final saturation, S_{sf}, on the solvent–water fractional flow curve, determined by the fractional flow of the injection itself. This intermediary saturation is found by mass balance considerations. The gas (i.e. solvent) velocity at the final saturation, S_{sf}, must be equivalent to the oil velocity at the intermediary saturation, S_{wf}. Formally, the line defined by Eq. (2.3.27) is extended to the oil–water fractional flow curve. Their

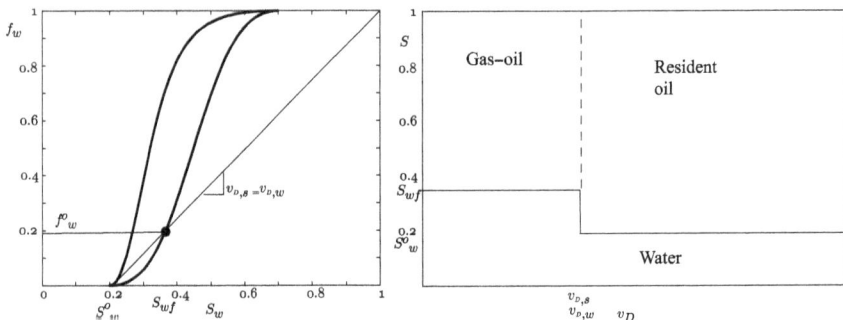

Figure 2.3.14. Showing the solvent and water injection velocities at an injection fractional flow where the solvent velocity is equal to the injected water velocity.

point of intersection determines the saturation of the leading water shock front,

$$v_{D,s} = \frac{1 - f_w^o}{1 - S_{sf}} = \frac{f_w^o - f_{wf}}{S_{sf} - S_{wf}}. \qquad (2.3.29)$$

The velocity of the leading shock front is given by the slope of the line from the initial saturation to (S_{wf}, f_{wf}). Again, in this case the saturation change is entirely within the shock front and there is no subsequent spreading wave.

Finally, the system with injection ratio f_w^o shown in Fig. 2.3.14 is the so-called optimum WAG ratio because it is the fractional flow for which the water and solvent fronts will move at the same velocity.

2.3.4.   *Three-Phase Flow*

There are many cases in EOR where flow will involve three phases, oil gas and water. Thus extensions of the flow analysis presented above to three-phase flow is also useful.

2.3.4.1.   *Fractional flow theory applied to three-phase flow*

Consider simultaneous flow of an oil, gas and aqueous phase that will be denoted with subscripts o, g and w, respectively. The saturation of the phases are related by,

$$S_o = 1 - S_w - S_g. \qquad (2.3.30)$$

The definitions of the fractional flow of each phase in the absence of capillary pressure gradients and gravity are given by,

$$f_w = \frac{\lambda_w}{\lambda_t}, \tag{2.3.31}$$

$$f_g = \frac{\lambda_g}{\lambda_t}. \tag{2.3.32}$$

The mass balance equations in terms of the fractional flow of each phase can be written as

$$\frac{\partial S_w}{\partial t_D} + \frac{\partial f_w}{\partial x_D} = 0, \tag{2.3.33}$$

$$\frac{\partial S_g}{\partial t_D} + \frac{\partial f_g}{\partial x_D} = 0. \tag{2.3.34}$$

The fractional flows, however, are functions of two variables (S_g and S_w) and thus expansion of Eq. (2.3.33) and (2.3.34) is given by,

$$\frac{\partial S_w}{\partial t_D} + \frac{\partial f_w}{\partial S_w}\frac{\partial S_w}{\partial x_D} + \frac{\partial f_w}{\partial S_g}\frac{\partial S_g}{\partial x_D} = 0, \tag{2.3.35}$$

$$\frac{\partial S_g}{\partial t_D} + \frac{\partial f_g}{\partial S_w}\frac{\partial S_w}{\partial x_D} + \frac{\partial f_g}{\partial S_g}\frac{\partial S_g}{\partial x_D} = 0. \tag{2.3.36}$$

Assuming that there are specific wave speeds, v_D, associated with saturations, as with the two-phase displacement of the Buckley–Leverett approach, results in the quadratic partial differential equation

$$\left(v_D - \frac{\partial f_w}{\partial S_w}\right)\left(v_D - \frac{\partial f_g}{\partial S_g}\right) - \frac{\partial f_w}{\partial S_g}\frac{\partial f_g}{\partial S_w} = 0. \tag{2.3.37}$$

For a given water and gas saturation, there are two values of v_D that satisfy Eq. (2.3.37).

2.3.4.2. *Models for three-phase relative permeability*

As with two-phase flow, the description of three-phase flow is sensitive to the relative permeability curves through the fractional flow functions. In the case of three-phase flow, however, relative permeability is not well understood from a theoretical standpoint and there is generally very little experimental data from which to draw empirical correlations. Summaries of the major issues are included in (Blunt, 2000; Juanes and Patzek, 2004; Baker, 1988; Delshad and Pope, 1989). This is an area of ongoing research but in summary, models currently considered the most accurate for representing three-phase relative permeability interpolate between values of two-phase relative permeabilities. In other words, for the oil relative permeability an interpolation is made between the relative permeability in the oil–water system, $k_{r,o(w)}$ and in the oil–gas system, $k_{r,o(g)}$, with the interpolation weighted by the water and gas saturations. The equivalent can be done for the water and gas relative permeabilities (Blunt, 2000). Following the model proposed by (Baker, 1988), the relative permeabilities for oil, water and gas can be given as

$$k_{r,o} = \frac{(S_w - S_{w,i})k_{r,o(w)} + (S_g - S_{g,r})k_{r,o(g)}}{(S_w - S_{w,i}) + (S_g - S_{g,r})}, \qquad (2.3.38)$$

$$k_{r,w} = \frac{(S_o - S_{o,i})k_{r,w(o)} + (S_g - S_{g,r})k_{r,w(g)}}{(S_o - S_{o,i}) + (S_g - S_{g,r})}, \qquad (2.3.39)$$

$$k_{r,g} = \frac{(S_w - S_{w,i})k_{r,g(w)} + (S_o - S_{o,i})k_{r,g(o)}}{(S_w - S_{w,i}) + (S_o - S_{o,i})}. \qquad (2.3.40)$$

In this model, $S_{w,i}$ is the irreducible water saturation, $S_{g,r}$ is the residual gas saturation in an oil–water displacement and $S_{o,i}$ is the initial oil saturation in a gas–water displacement, the latter two are usually zero. Using this model, the three-phase relative permeability functions are obtained through the measurement of six two-phase relative permeability curves, both phases in an oil–water, oil–gas and gas–water displacement.

2.4.   Chemical EOR: Polymers and Low Salinity Flooding

Chemical EOR techniques involve changing the chemistry of the injection water either by adding chemical components (polymers, Alkali–Surfactant–Polymer (ASP), sulphate salts) or removing them (low salinity water flooding) in order to improve oil recovery. Of these techniques, polymer flooding has the longest history (dating back to the late 1960s) and is the only one to have been deployed at scale in the field. ASP has been discussed for almost as long as polymer flooding although it has only been used in field pilots to date. It was devised to combine the benefits of using a surfactant to reduce the oil–water interfacial tension and hence reduce capillary trapping of the oil and a polymer to provide mobility control whilst including an alkali to prevent both of these being lost by adsorption onto the rock. In contrast, low salinity water injection has only been extensively discussed since it was rediscovered by (Yildiz and Morrow, 1996). There have been several single well field tests as well as one interwell field trial (Seccombe *et al.*, 2010) but it will not be deliberately deployed until Clair Ridge comes on stream in 2017. High sulphate water injection is a specialised EOR technique for chalk reservoirs e.g. (Zhang *et al.*, 2007). Only polymer flooding and low salinity water flooding will be discussed in detail in this section.

2.4.1. *Polymer Flooding*

In polymer flooding, water soluble polymers are added to the injection water to viscosify the water. It is usually used for the recovery of more viscous, but still mobile oils (up to a viscosity of approximately 0.1 Pa·s) in higher permeability reservoirs (several hundred mD or higher). Using a polymer solution increases the mobility ratio and thus reduces the tendency for the water to bypass the oil through either viscous fingering or channelling. Additional oil may also be recovered by increased viscous cross-flow (Sorbie, 1991) thus polymer flooding is the only one of the EOR processes targeted at improving macroscopic sweep as well as improving microscopic displacement efficiency. One of the main disadvantages of this process, apart from cost, is that injection rates tend to be lower because of the increased water viscosity, thus overall production rate may be reduced. Nonetheless, oil production rates may be higher because of the reduced water cut. Some polymers are shear thinning, meaning that their viscosity reduces at high shear rates (such as those that occur in and near the wellbore). In this case, there will be a smaller reduction in injectivity. Further problems may arise if the polymer is adsorbed onto the rock or gets trapped in the pore throats. The first effect will tend to reduce the effectiveness of the polymer (as there will be less in solution and thus the solution viscosity will not be as high as planned) whilst the second effect may result in permeability reduction and further loss of injectivity.

The principal chemicals used are polyacrylamide (used in 95% of projects in the literature) and xanthan gum e.g. (Morel *et al.*, 2008). Polyacrylamide is the cheaper of the two and does not degrade easily. This latter attribute also means that it persists in the natural environment and so it cannot be applied in Norwegian oil fields unless the produced water is reinjected. It is also less stable in higher salinity reservoir waters. Xanthan gum is more stable at these higher salinities but degrades too easily for most reservoir uses. It is also more expensive. Neither of these chemicals is stable at very high temperatures (i.e. deep reservoirs) or in carbonate reservoirs where the connate water is particularly hard. Current research is focussed on developing more robust, cheap and environmentally friendly alternatives.

At the time of writing, China is the only country in the world where there has been and continues to be large scale oil production from polymer flooding. This is primarily a result of project economics. Field scale deployment of polymer flooding needs large volumes of chemicals (e.g. using xanthan in the proposed Dalia development, offshore Angola, would have required a 50% increase in world production of xanthan, (Morel *et al.*, 2008)), transportation of those chemicals and then the facilities to mix and inject those chemicals without breaking down the molecules and thus reducing their viscosifying effect. Nonetheless in 2004 it was reported that 250,000 barrels of oil a day were being produced as a result of polymer injection into the Daqing field in China (Chang *et al.*, 2006). The high oil price until 2014 combined with the increased focus on maintaining production from mature fields led to renewed interest in polymer flooding and tests have begun on commercial scales using both polymer floods and ASP flooding in Oman (Stoll *et al.*, 2011; Al-Saadi *et al.*, 2012). It has also been identified as one of the three highest potential EOR technologies for the North Sea (McCormack *et al.*, 2014).

Fractional flow analysis can be used to understand how polymer flooding increases the microscopic displacement efficiency. This analysis is similar to the miscible gas injection process described above with two key differences: first, the dissolved chemical concentration, C_1, of interest is polymer and it is dissolved in the aqueous phase. Second, one of the key processes to capture in polymer flooding is the reversible adsorption of the polymer onto solid walls of the rock pores. This is done by accounting for some amount of the concentration, C_1, as made up of the adsorbed concentration, c_{1a}, the amount of adsorbed polymer per unit pore volume. Figure 2.4.1 shows two fractional flow curves representing the displacement processes between the water–oil system and the aqueous polymer solution–oil system, with the mobility lower in the polymer solution displacement due to the increased viscosity. In this iteration thus the total concentration, including the adsorbed component, is given by,

$$C_1 = c_{1w}S_w + c_{1o}S_o + c_{1a} = c_{1w}S_w + c_{1a}.$$

The total flux is given by,

$$F_1 = c_{1w}f_w + c_{1o}f_o = c_{1w}f_w.$$

It is usually assumed that there exists a simple relationship between the concentration of polymer in the aqueous phase and the adsorbed concentration, e.g.

$$c_{1a} = Rc_{1w},$$

where R is a proportionality constant known as the *retardation factor*. This is because processes of reversible adsorption have the

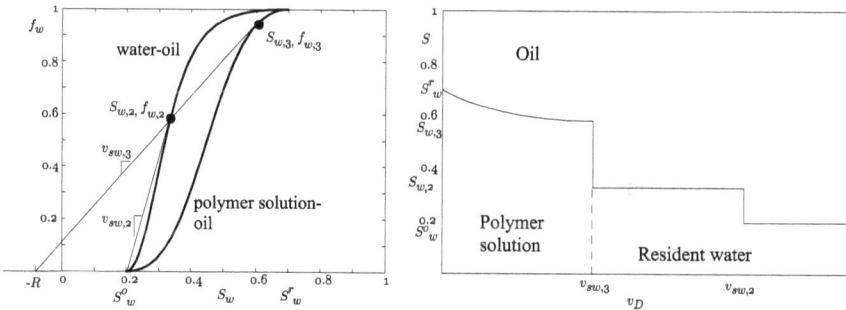

Figure 2.4.1. Fractional flow analysis for polymer flooding.

ultimate effect of slowing down a concentration wave as will be seen below.

From this, the mass balance Eq. (2.3.19) becomes

$$\frac{\partial[c_{1w}S_w + c_{1a}]}{\partial t_D} + \frac{\partial[c_{1w}(f_w)]}{\partial x_d} = 0$$

$$= S_w\frac{\partial c_{1w}}{\partial t_D} + c_{1w}\frac{\partial S_w}{\partial t_D} + \frac{\partial c_{1a}}{\partial t_D} + f_w\frac{\partial c_{1w}}{\partial x_D} + c_{1w}\frac{\partial f_w}{\partial x_D}$$

$$= S_w\frac{\partial c_{1w}}{\partial t_D} + \frac{\partial c_{1a}}{\partial c_{1w}}\frac{\partial c_{1w}}{\partial t_D} + f_w\frac{\partial c_{1w}}{\partial x_D}$$

$$= (S_w + R)\frac{\partial c_{1w}}{\partial t_D} + f_w\frac{\partial c_{1w}}{\partial x_D}$$

$$= \frac{\partial c_{1w}}{\partial t_D} + \frac{f_w}{S_w + R}\frac{\partial c_{1w}}{\partial x_D} = 0.$$

The velocity of the concentration of the polymer solution is thus

$$v_{sw} = \frac{f_w}{S_w + R}.$$

Consider a polymer flood where the aqueous polymer solution is injected at a fractional flow of 1 (Fig. 2.4.1). There will be two shock fronts: A leading shock front where resident brine is banked and a second shock front between the polymer solution and the resident brine. The particular saturation, fractional flow and velocities of those fronts are found through mass balance considerations. In particular, the shock saturation for the arrival of the polymer food is given by equating,

$$\frac{f_w}{S_{w,3} + R} = \left.\frac{df_w}{dS_w}\right|_{S_w = S_{w,3}}.$$

Similarly, the saturation front of the resident water bank is given by the following equation further to the concentration shock velocity between the polymer solution and the resident brine,

$$\frac{f_{w,3}}{S_{w,3} + R} = \frac{f_{w,3} - f_{w,2}}{S_{w,3} - S_{w,2}}.$$

Finally there is a rarefaction of the saturation wave to the final residual oil saturation.

2.4.2. *Low Salinity Water Flooding*

Low salinity water flooding involves the injection of water with a reduced salt concentration. It appears to work by making the reservoir rock more water-wet, reducing the residual oil saturation and changing the fractional flow behaviour. It is therefore targeted at improving microscopic displacement efficiency. In principal, it is one of the cheaper and simpler chemical EOR techniques as it only needs the installation of desalination plants rather than the injection of large volumes of expensive chemicals. There is still some debate about the exact mechanism causing this wettability alteration (e.g. Lager *et al.*, 2008; Tang and Morrow, 1999) and indeed some authors (Thyne, 2011; Skrettingland *et al.*, 2011) question whether there is any improved oil recovery at all, resulting from this process. Nonetheless, most of the literature suggests that there will be improved oil recovery (e.g. Webb *et al.*, 2003; Mahani *et al.*, 2011).

There appear to be four criteria necessary for low salinity water-flooding to be a success: (a) the crude oil should contain a significant proportion of polar components (especially organic acids), (b) the connate water should contain divalent cations (especially Ca^{2+} and Mg^{2+}), (c) the reservoir rock must be sandstone, ideally with a proportion of clays, especially kaolinite and (d) the injected water must be softened, i.e. has a reduced divalent cation content. These criteria are most consistent with the multi-component ion exchange mechanism for low salinity waterflooding EOR as originally proposed by (Lager *et al.*, 2008). More recent research (e.g. Hassenkam *et al.*, 2011) also tends to support this model but it is possible that the release and migration of clay fines may also play a role in some rocks (Tang and Morrow, 1999).

To date there have been no deliberate field wide implementations of low salinity water injection although there have been some inadvertent deployments (e.g. in the Omar field in Syria, (Mahani *et al.*, 2011)). There have been a number of single well tracer tests and one well-to-well pilot (Webb *et al.*, 2003). The first planned low salinity development will be at Clair Ridge in the North Sea with production due to start in 2017 (McCormack *et al.*, 2014) and, along with polymer flooding and gas injection, it is one of the three

EOR techniques with most potential for improving oil recovery in the North Sea as identified by (McCormack *et al.*, 2014). The main practical factor limiting its deployment is the availability of space for the desalination plants on ageing platforms.

The simplest model for predicting the performance of low salinity waterfloods is that of (Jerauld *et al.*, 2008). In this model, there are two sets of relative permeability curves, a more oil wet pair of curves corresponding to a high salinity water-flood and a more water-wet pair, corresponding to the low salinity waterflood. The reservoir originally contains oil and high salinity connate water. Lower salinity water is injected displacing both the high salinity water and some of the oil. This results in a 1D displacement formed of two shocks and a rarefaction, similar to that seen when performing the fractional flow analysis for polymer flooding shown in the previous section. As in the polymer flooding case there is a connate water bank ahead of the low salinity front.

As in the polymer flooding case we draw a graph showing the fractional flow curve for the high salinity water injection relative permeabilities and the fractional flow curve resulting from using the low salinity water injection relative permeabilities. In this case, we draw a line from the point (0, 0) (see Fig. 2.4.2) that is tangent to the low salinity fractional flow curve. This is because the low salinity and high salinity water are fully miscible and we assume there is no retardation of the low salinity water. The point of intersection

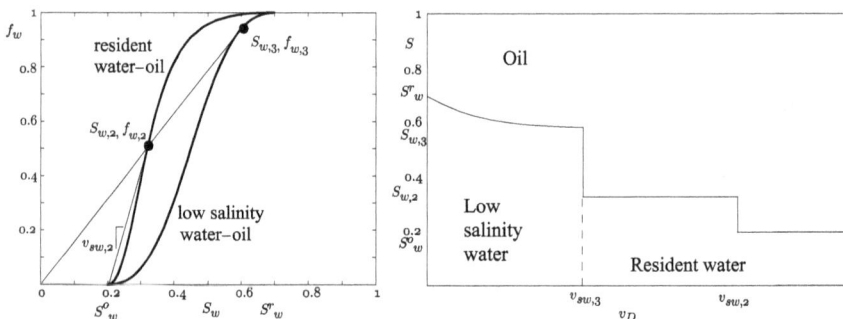

Figure 2.4.2. Fractional flow analysis for low salinity water flooding applied for primary recovery.

of this line with the high salinity fractional flow curve $(S_{w,2}, f_{w,2})$ gives the shock front saturation of the leading edge of the connate water bank $(S_{w,2})$ and the point of tangency gives the shock front saturation of the leading edge of the low salinity water $(S_{w,3})$. The characteristic velocity of the leading shock is determined by drawing a line from the connate water saturation of the high salinity fractional flow to the point $(S_{w,2}, f_{w,2})$ and calculating the gradient of that line. The characteristic velocity of the trailing shock is simply the gradient of the original tangent line. Finally, there is a rarefaction to the residual oil saturation. If low salinity flooding is applied as a secondary or tertiary recovery process, the oil saturation will be low. This will result in the creation of an oil bank, characteristic of enhanced recovery induced by a wetting state change (Fig. 2.4.3).

Jerauld *et al.* (2008) proposed a refinement of this model when implementing it in a numerical simulator. They asserted that the injected low salinity water would only begin to alter oil recovery when the water salinity was lower than a specified upper salinity threshold (e.g. 7,000 ppm) and that it would only be fully effective once salinity was below a lower threshold (e.g. 1,000 ppm). For salinities below the lower threshold then the flow would be described by the more water-wet relative permeabilities whilst for salinities above the upper threshold the flow would be described by the more oil-wet relative permeabilities. For intermediate salinities then the relative permeabilities would be a linear combination of the high

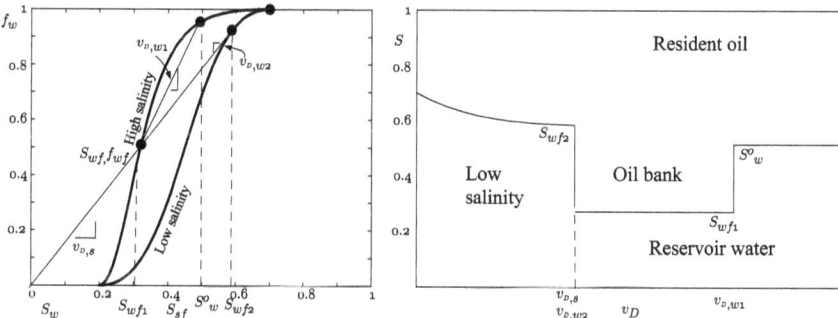

Figure 2.4.3. Fractional flow analysis for low salinity flooding applied as a secondary or tertiary recovery mechanism.

and low salinity relative permeabilities. This allows the approximate modelling of the effect of mixing between the injected low salinity water and the high salinity connate water on oil recovery. More sophisticated models (such as those of Dang *et al.* (2013)) use a more sophisticated ion exchange model to determine which relative permeability curves should be used during the displacement.

2.5.   Practical Considerations

In Sec. 2.1, we saw that EOR contributed only 2% to global oil production in 2014. Current predictions suggest that it will still contribute less than 10% by 2030 (Aleklett *et al.*, 2010). We have seen that the average recovery factor from oil fields around the world is between 30% and 40% although in some fields such as Forties in the North Sea it is believed the recovery factor has reached 65%. We also know that it is becoming increasingly difficult to find new oil fields that contain significant volumes of conventional oil. Instead the world is looking increasingly towards the exploitation of unconventional oil and gas reserves such as so called shale oil and gas and the very viscous bitumens found in Canada and Venezuela. Given the variety of EOR techniques available to reservoir engineers and the fact that the majority of these have been discussed, researched and often piloted in the field for more than 40 years, why is the contribution of EOR to world production so low and projected to grow so little?

The reality is that, although it is technically feasible to significantly improve oil recovery using EOR techniques, it is often not economically feasible to do so. EOR requires very large volumes of gas or chemicals. These injectants are more expensive than water and the benefits of using them may not be seen for many years. This is one reason why national oil companies (e.g. in China) are more likely to implement EOR than independent oil companies. Governments may take a longer term view, husbanding their nation's resources for the benefit of the country whilst independent oil companies need to pay dividends to their shareholders every year.

Sometimes there are practical considerations limiting the deployment of EOR e.g. there is simply not enough hydrocarbon gas available to replace the produced oil and maintain reservoir pressure above the minimum miscibility pressure and it is not economic to build pipelines to import that gas from elsewhere. We also know that hydrocarbon gas in itself is a valuable commodity. There is therefore a commercial decision to be made balancing the immediate income from gas sales versus the additional oil that may result some years after injecting the gas into the reservoir. Most successful miscible gas injection projects have been in fields producing significant quantities of gas that cannot easily be sold because the field is very remote from centres of population that could use the gas (e.g. Prudhoe Bay in Alaska) or because there is no export route for the produced gas (e.g. Ula in the North Sea).

From the chemical perspective we have seen, for example, that using xanthan as the polymer in the Dalia field would have required a 50% increase in world production of this chemical (Morel *et al.*, 2008). In general it is not economic to build whole new factories to supply this volume of polymer. In addition, many oil fields are in geographically remote regions of the world, often in extreme climates (e.g. the Arctic, or the desert). In the Arctic, it may only be possible to transport materials to the well head for the 6 months of the year when the ground is frozen. Transporting large volumes of polymer (even if they were available) to the middle of the desert may require so many trucks that it may not be practical. Even if it were, there may not be sufficient good quality water available in which to dissolve it prior to injection. This is also a factor that limits the potential deployment of low salinity water injection.

It is, of course, also important to bear in mind environmental considerations when planning an EOR scheme. The chief concern in most schemes is the disposal of produced water. Even ordinary produced water needs to be reinjected or treated before it can be disposed of on the surface. This treatment becomes more important when chemicals, such as polymers or surfactants, are injected in the solution. We have already noted that polyacrylamide cannot be injected into Norwegian fields because it persists in the environment.

Most, if not all, EOR techniques are most effective if applied as secondary recovery schemes. This is because water will tend to trap oil in pores and bypass oil in lower permeability zones. EOR injectant applied after a waterflood may then simply follow the flow paths already established by the injected water so less oil will be accessed. It will also take longer for the oil that is swept up by the EOR injectant to reach the production well. Nonetheless, many engineers will only consider using EOR after a waterflood. This may be because at the time the field was discovered the oil price was too low to justify the additional expense of an EOR scheme. However, even when the oil price has been higher many engineers have preferred to start production using water injection. In this case, they reason that this will give them a better understanding of the geological heterogeneity in their reservoir and will thus reduce the risk of a poor outcome when they finally deploy EOR. It is not always clear that this reduction in perceived risk outweighs the loss of additional recovery from applying the EOR scheme in secondary mode. The response from tertiary EOR schemes may be so slow that it can take a year or more for there to be a noticeable slowing of the decline in oil rate or indeed an observable improvement in oil rate. As an example it took more than two years from the start of gas injection in Magnus to see a noticeable improvement in oil recovery (Brodie *et al.*, 2012). It can also be difficult to clearly relate this improvement in oil rate to the EOR process rather than more recent activities such as well workovers. In any case, this delay may make the scheme uneconomic.

If it is likely that EOR will be applied post-waterflood then it is important to plan the field infrastructure to allow for this. If this is not done then it may become prohibitively expensive to retrofit these facilities e.g. many offshore platforms on older fields in the North Sea have no space for the installation of desalination plants or the preparation of polymer solutions. Injection of carbon dioxide, whilst potentially effective in terms of recovering additional oil and also providing a means of mitigating climate change through the storage of the gas, would require installation of specialised pipework and wells that are corrosion resistant. To give an indication of the potential costs, in 2014 drilling a new well in the North Sea could cost £100 million. In the case of CO_2 EOR/sequestration there would also be the additional cost of a subsea pipeline network to distribute the CO_2 to fields from the power stations where it would be captured.

Leaving aside economic considerations, the appraisal, planning and development of EOR schemes can be significantly more complex than for a conventional water flood. We have already seen in the previous chapters how 1D fractional flow analysis for WAG, polymer and low salinity water injection is more involved than for water flooding, not to mention the additional understanding of phase behaviour that is needed when planning a miscible gas injection scheme. Leaving aside the logistical considerations discussed in preceding paragraphs and the additional facilities design required, further specialised laboratory tests may be needed to determine EOR specific data such as the parameters needed to describe phase behaviour when injecting a miscible or multi-contact miscible gas, the relative permeability curves needed to plan a low salinity flood in the reservoir of interest or to quantify the expected levels of polymer adsorption. Having estimated the microscopic displacement efficiency of possible EOR schemes using the analytical techniques described in previous chapters, it will then be necessary to assess the impact of geological heterogeneity on macroscopic sweep as well as the range of possible outcomes due to the uncertainty in that heterogeneity. This requires the use of specialised reservoir simulation software which in turn relies on specialised reservoir engineering expertise (e.g. chemistry of rock–fluid interactions for low salinity waterflooding, chemistry and flow behaviour of the emulsions that form during ASP flooding, phase behaviour of multi-contact miscible displacements and use of equations of state). In some cases this additional complexity may be sufficient to dissuade engineers and their managers from even considering an EOR scheme.

Despite these various economic, practical and technical difficulties EOR still has the potential to significantly improve oil recovery. McCormack *et al.* (2014) suggest that EOR could recover an additional 0.6–1.2 billion barrels of oil, from the UKCS alone. Worldwide it could increase the reserves to production ratio from 40 years to perhaps 45 or 50 years. This can only be achieved, however, through a combination of good reservoir engineering, longer term planning, innovative research and, of course, a sufficiently high oil price to make these schemes economic. Burning oil in the form of petrol and diesel does increase the amount of CO_2 in the atmosphere and is thus one of the cause of anthropogenic climate change, but at the time of writing this chapter there is no viable alternative fuel for transportation that can be immediately be applied worldwide in all forms of transport. Until this becomes available crude oil will still be needed and EOR has the potential to help supply that demand.

References

Al-Saadi, F. S., Amri, B. A., Nofli, S., Wunnik, J. V., Jaspers, H. F., Harthi, S., Shuaili, K., Cherukupalli, P. K. and Chakravarthi, R. (2012). Polymer flooding in a large field in South Oman — initial results and future plans. *SPE EOR Conference at Oil and Gas West Asia* held in Muscat, Oman, (SPE 154665).

Aleklett, K., Hook, M., Jakobsson, K., Lardelli, M., Snowden, S. and Soderbergh, B. (2010). The peak of the oil age — analyzing the world oil production reference scenario in World Energy Outlook 2008. *Energ. Policy*, **38**(3), 1398–1414.

Anderson, W. (1987). Wettability literature survey — part 6: The effects of wettability on waterflooding. *J. Petrol. Technol.*, **39**(12), 1605–1622.

ARI (2009). CO_2 storage in depleted oilfields: Global application criteria for carbon dioxide enhanced oil recovery. Technical Report 2009–12, IEA Greenhouse Gas Program.

Baker, L. E. (1988). Three phase relative permeability correlations. *SPE/DOE Enhanced Oil Recovery Symposium* Tulsa, Oklahoma, (SPE/DOE 17369).

Blunt, M. J. (2000). An empirical model for three-phase relative permeability. *Soc. Petrol. Eng. J.*, **5**(4), 435–445.

Blunt, M. J. (2013). *Reservoir Performance Predictors Course Notes*. Imperial College London.

Brodie, J., Jhaveri, B., Moulds, T. and Mellemstrand, H. S. (2012). Review of gas injection projects in BP. In *Proceedings of the 18th SPE Improved Oil Recovery Symposium*, Tulsa, OK 14–18 April.

Chang, H., Zhang, Z., Wang, Q., Zu, Z., Guo, Z., Sun, H., Cao, X. and Qiao, Q. (2006). Advances in polymer flooding and alkaline/surfactant/polymer processes as developed and applied in the People's Republic of China. *J. Petrol. Technol.*, **58**(2), 84–91.

Christie, M., Jones, A. and Muggeridge, A. (1990). Comparison between laboratory experiments and detailed simulations of unstable miscible displacement influenced by gravity. *North Sea Oil and Gas Reservoirs-II*.

Dang, C., Nghiem, L. X., Chen, Z. and Nguyen, Q. P. (2013). Modeling low salinity waterflooding: Ion exchange, geochemistry and wettability alteration. *SPE Annual Technical Conference and Exhibition*, New Orleans, Louisiana, USA, 30 September–2 October, (SPE 166447).

Delshad, M. and Pope, G. A. (1989). Comparison of the three-phase oil relative permeability models. *Transport in Porous Media*, **4**, 59–83.

Dietz, D. (1953). A theoretical approach to the problem of encroaching and bypassing edge water. *Pro. Koninkl. Ned. Akad. Wetemchap.*, S36.

Djabbarov, S. (2014). Experimental and numerical studies of first contact miscible injection in a quarter five spot pattern. Master's thesis, Imperial College London.

Dumore, J. (1964). Stability considerations in downward miscible displacements. *Soc. Petrol. Eng. J.*, **4**(4), 356–362.

Dykstra, H. and Parsons, R. (1950). The prediction of oil recovery by water-flood. *Secondary Recovery of Oil in the United States*, 2nd ed. Dallas: API, pp. 160–174.

Fayers, F. J. and Muggeridge, A. H. (1990). Extensions to Dietz theory and behavior of gravity tongues in slightly tilted reservoirs. *SPE Reservoir Eng.*, **5**(04), 487–494.

Gardner, J. W., Orr, F. M. Jr. and Patel, P. (1981). The effect of phase behavior on CO_2-flood displacement efficiency. *J. Petrol. Technol.*, **33**(11), 2061–2081.

Hassenkam, T., Pedersen, C., Dalby, K., Austad, T. and Stipp, S. (2011). Pore scale observation of low salinity effects on outcrop and oil reservoir sandstone. *Colloid Surface A*, **390**, 179–188.

Hite, J., Stosur, G., Carnaham, N. and Miller, K. (2003). Guest editorial. ior and eor: effective communication requires a definition of terms. *J. Petrol. Technol.*, **55**(16).

Homsy, G. (1987). Viscous fingering in porous media. *Annu. Rev. Fluid Mech.*, **19**, 271–311.

IEA (2013a). *Resources to Reserves 2013 — Oil, Gas and Coal Technologies for the Energy Markets of the Future*. International Energy Agency.

IEA (2013b). *World Energy Outlook 2013*. International Energy Agency.

Jensen, J. and Currie, I. (1990). A new method for estimating the Dykstra-Parsons coefficient to characterize reservoir heterogeneity. *SPE Reservoir Eng.*, **5**(3), 369–374.

Jerauld, G. R., Lin, C. Y., Webb, K. J. and Seccombe, J. C. (2008). Modeling low-salinity waterflooding. *SPE Reserv. Eval. Eng.*, **11**(6), 1000–1012.

Juanes, R. and Patzek, T. W. (2004). Three-phase displacement theory: An improved description of relative permeabilities. *Soc. Petrol. Eng. J.*, **9**(3), 1–12.

Koottungal, L. (2014). 2014 worldwide EOR survey. *Oil and Gas J.*, 04/07/2014.

Koval, E. (1963). A method for predicting the performance of unstable miscible displacements in heterogeneous media. *Soc. Petrol. Eng. J.*, **3**(02), 145–154.

Lager, A., Webb, K. J., Collins, I. R. and Richmond, D. M. (2008). LoSal enhanced oil recovery: Evidence of enhanced oil recovery at the reservoir scale. *SPE Symposium on Improved Oil Recovery*, Tulsa, Oklahoma, USA, 20–23 April (SPE-113976-MS).

Lake, L. W. (1989). *Enhanced Oil Recovery.* Society of Petroleum Engineers, Texas, United States.

Lenormand, R., Zarcone, C. and Sarr, A. (1983). Mechanism of the displacement of one fluid by another in a network of capillary ducts. *J. Fluid Mech.*, **135**, 337–353.

Mahadevan, J., Lake, L. W. and Johns, R. T. (2003). Estimation of true dispersivity in field-scale permeable media. *Soc. Petrol. Eng. J.*, **8**(3), 272–279.

Mahani, H., Sorop, T. G., Ligthelm, D., Brooks, A. D., Vledder, P., Mozahem, F. and Ali, Y. (2011). Analysis of field responses to low salinity waterflooding in secondary and tertiary mode in syria. *SPE Europec/EAGE Annual Conference and Exhibition*, Vienna, Austria, (SPE 142960).

McCormack, M. P., Thomas, J. M. and Mackie, K. (2014). Maximising enhanced oil recovery opportunities in ukcs through collaboration. *Abu Dhabi International Petroleum Exhibition and Conference*, (SPE 172017).

Morel, D., Vert, M., Jouenne, S. and Nahas, E. (2008). Polymer injection in deep offshore field: The dalia case. *SPE Annual Technical Conference and Exhibition*, Denver, Colorado, September, (SPE 116672).

Muggeridge, A., Cockin, A., Webb, K., Frampton, H., Collins, I., Moulds, T. and Salino, P. (2012). Recovery rates, enhanced oil recovery and technological limits. *Philos. T. Roy. Soc. A*, **372**(20120320).

Orr, F. M. Jr., Yu, A. D. and Lien, C. L. (1981). Phase behavior of co2 and crude oil in low-temperature reservoirs. *Soc. Petrol. Eng. J.*, **21**(4), 480–492.

Orr, F. M. Jr. (2007). *Theory of Gas Injection Processes.* Tie-Line Publications.

Pentland, C. H., Itsekiri, E., Mansoori, S. K. A. and Iglauer, S. (2010). Measurement of nonwetting-phase trapping in sandpacks. *Soc. Petrol. Eng. J.*, **15**(02), 274–281.

Pope, G. A. (1980). The application of fractional flow theory to enhanced oil recovery. *Soc. Petrol. Eng. J.*, **20**(03).

Roof, J. (1970). Snap-off of oil droplets in water-wet pores. *Soc. Petrol. Eng. J.*, **10**(01), 85–91.

Salathiel, R. (1973). Oil recovery by surface film drainage in mixed-wettability rocks. *J. Petrol. Technol.*, **25**(10), 1216–1224.

Schmalz, J. P. and Rahme, H. D. (1950). The variation of waterflood performance with variation in permeability profile. *Producers Monthly*, **15**(9), 9–12.

Seccombe, J., Lager, A., Jerauld, G., Jhaveri, B., Buikema, T., Bassler, S., Denis, J., Webb, K., Cockin, A. and Fueg, E. (2010). Demonstration of low-salinity EOR at interwell scale, Endicott field, Alaska. *SPE Improved Oil Recovery Symposium*, Tulsa, Oklahoma, 24–28 April, (SPE-129692-MS).

Seright, R. S., Lane, R. H. and Sydansk, R. D. (2003). A strategy for attacking excess water production. *SPE Prod. Facil.*, **18**, 158–169.

Skrettingland, K., Holt, T., Tweheyo, M. T. and Skjevark, I. (2011). Snorre low-salinity-water injection-coreflooding experiments and single-well field pilot. *SPE Reserv. Eval. Eng.*, **14**(2), 182–192.

Smalley, P. C., Ross, B., Brown, C. E., Moulds, T. P. and Smith, M. J. (2009). Reservoir technical limits: A framework for maximizing recovery from oil fields. *SPE Reserv. Eval. Eng.*, **12**(04), 610–617.

Sorbie, K. S. (1991). *Polymer-Improved Oil Recovery.* Springer.

Sorbie, K. S. and Seright, R. S. (1992). Gel placement in heterogeneous systems with crossflow. *Proceedings SPE/DOE Enhanced Oil Recovery Symposium,* Tulsa, OK, (SPE-24192-MS).

Stoll, W., al Shureqi, H., Finol, J., Al-Harthy, S., Oyemade, S., de Kruijf, A., van Wunnik, J., Arkesteijn, F., Bouwmeester, R. and Faber, M. (2011). Alkaline/surfactant/polymer flood: From the laboratory to the field. *SPE Reserv. Eval. Eng.*, **14**(06), 702–712.

Taber, J., Martin, F. and Seright, R. (1997a). Eor screening criteria revisited — part 1: Introduction to screening criteria and enhanced recovery field projects. *SPE Reservoir Eng.*, **12**(3), 189–198.

Taber, J., Martin, F. and Seright, R. (1997b). EOR screening criteria revisited — part 2: Applications and impact of oil prices. *SPE Reservoir Eng.*, **12**(03), 199–205.

Tang, G. Q. and Morrow, N. R. (1999). Salinity, temperature, oil composition and oil recovery by waterflooding. *SPE Reservoir Eng.*, **12**, 269–276.

Tchelepi, H. A. and Orr, F.M. Jr. (1994). Interaction of viscous fingering, permeability heterogeneity, and gravity segregation in three dimensions. *SPE Reservoir Eng.*, **9**(04), 266–271.

Thyne, G. (2011). Evaluation of the effect of low salinity waterflooding for 26 fields in Wyoming. *SPE Annual Technical Conference and Exhibition,* Denver, Colorado, (SPE 147410).

Todd, M. and Longstaff, W. (1972). The development, testing and application of a numerical simulator for predicting miscible flood performance. *J. Petrol. Technol.*, **24**(7), 874–882.

USEIA (2014). Annual Energy Outlook 2014 with projections to 2014. Technical Report DOE/EIA-0383(2014), US Energy Information Administration.

Wallace, M. and Kuuskraa, V. (2014). Near-term projections of CO_2 utilization for enhanced oil recovery. Technical Report DOE/NETL-2014/1648, National Energy Technology Laboratory.

Webb, K. J., Black, C. J. J. and Al-Ajeel, H. (2003). Low salinity oil recovery- log inject log. *Middle East Oil Show,* Bahrain, 9–12 June, (SPE-81460-MS).

Yildiz, H. O. and Morrow, N. R. (1996). Effect of brine composition on recovery of Moutray crude oil by waterflooding. *J. Petrol. Sci. Eng.*, **14**, 159–168.

Zhang, P. M., Tweheyo, M. T. and Austad, T. (2007). Wettability alteration and improved oil recovery by spontaneous imbibition of seawater into chalk: impact of the potential determining ions Ca^{2+}, Mg^{2+} and SO_2^{-4}. *Colloid Surface A,* **301**, 199–208.

Chapter 3

Numerical Simulation

Dave Waldren

Petroleum Consulting and Training Ltd.
Cobham, Surrey, UK

3.1. Reservoir Models

3.1.1. *Introduction*

The purpose of modelling, in any discipline, is to be able to simulate a process or physical phenomenon in a controlled environment in order to monitor the response of the model to a coherent set of environmental changes.

Models may be broadly classified into the categories of

(1) theory based models; or
(2) phenomenological or parametric models.

Usually, phenomonological models are used because either no theory exists or the interaction between several phenomena leads to complexity in the theory whereas the overall behaviour may be easily parameterised.

The real gas law:

$$pV = nzRT. \qquad (3.1.1)$$

Using a temperature and pressure dependent correction factor, z, is a phenomenological modification of the ideal gas law which is a model based on the classical kinetic theory of gases.

An alternative classification of models is:

(1) physical models; or
(2) mathematical models.

Usually, the results of physical models are used in order to make a parametric or phenomenological description of the process being studied. Newton's laws of motion and Planck's law of radiation are two examples of laws derived by these methods.

The procedure of experimental observation and parametric analysis has been the basis for advance in many branches of the physical sciences since the time of alchemists and indeed, application of models can be said to have been the major factor in transforming disciplines from arts into sciences.

Although, in this course, interest is focused on mathematical models of hydrocarbon reservoirs, it is important to consider the physical models in order to understand their role in reservoir simulation.

3.1.2. *Physical Models*

Physical models may be divided into three categories:

(1) Analogue models;
(2) Comprehensive models; and
(3) Elemental models.

3.1.2.1. *Analogue models*

The basis of analogue models is the study of a phenomenon which has similar behaviour to the one of interest, in order to investigate the dependence on parameters which have analogues in the system of interest.

Examples of such models are:

(1) heat flow to simulate single phase fluid flow; and
(2) electrostatic experiments to simulate flux lines between wells.

These models are little used in reservoir simulation.

3.1.2.2.   *Comprehensive models*

Comprehensive models are those in which all the important facets of the reservoir production are included in a physical model. Because of the size of the problem most of these are scale models.

In scaled models, reservoir dimensions, rock properties, fluid properties and velocities must be scaled so that proportionally, the forces in the model conform to those in the reservoir. In practice this is difficult, if not impossible, to achieve as, for example, scaling down the permeability would usually have the effect of scaling up the capillary pressure hence altering the ratio between capillary, gravity and viscous forces. Such alteration would render the results of the model inappropriate for direct application to field scale problems.

An example of a comprehensive model is the vertical displacement of oil by gas in a long core at reservoir conditions. Here, the reason for performing the experiment is a lack of ability to adequately represent the interaction of the displacement and mass transfer effects.

Although these models are useful and the results can be included in reservoir simulation models, they have very little predictive power as the basic mechanisms are not parameterised.

3.1.2.3.   *Elemental models*

Elemental models are constructed as a microcosm of a certain aspect of the reservoir behaviour and as such the experiments performed on reservoir core plugs and fluid samples are examples of this genre. By their nature these experiments concentrate on a restricted range of phenomena actually occurring in the reservoir and attempt to isolate these from the possible interactions between the studied phenomena and other effects. Because of this requirement, these experiments must be carefully designed to specifically exclude these extraneous effects.

An example of such a model is the behaviour of oil and gas flowing in a porous medium giving the experimentally observed relative permeability. As the purpose of the experiment is to quantify the multi-phase flow characteristics, the experiment is designed to eliminate any effects of mass transfer between phases and gravitational segregation of fluids.

Depending on the fluid characteristics and production mechanism of the reservoir, this "elemental model" might or might not be directly applicable to the flow of fluids in the reservoir.

3.1.3. *Mathematical Models*

A mathematical model of a system is simply an equation which relates the behaviour of the system, expressed in terms of observable variables, to some parameters which describe the system. These equations are frequently described as physical "laws".

Apart from very few fundamental laws, all of these are generalisations with some restricted area of validity. Examples of such generalisations are Newton's laws of motion, the ideal gas law and Darcy's law.

As the physical complexity of any system increases, the restrictions imposed, in order to produce an analytically or easily soluble mathematical model, increase. The danger in using any mathematical model is that if it is used outside the range imposed by the assumptions inherent in its formulation, the results may appear unphysical or, worse, they may not appear unrealistic but might give a completely false picture of the important mechanisms and the overall behaviour.

Examples of mathematical models applied to petroleum reservoirs are decline curve analysis and material balance. These techniques are very useful in characterising well response in an unchanging production scheme and overall reservoir and aquifer behaviour but, because of the simplifying assumptions, they are of less use for detailed reservoir description purposes or for formulating remedial action for production problems.

3.1.4. *Numerical Models*

A more detailed mathematical model can be constructed by subdividing the reservoir into small volume elements and applying the laws of mass conservation and fluid flow to each element. By letting the elements tend to zero volume, the equations for movement of fluids in a porous medium can be constructed. For the formulation of these equations see the section on Equations and Terminology.

The resulting equations are nonlinear partial differential equations which are almost always too complicated to solve analytically even for a simple description of the hydrocarbon fluid properties.

Further approximations are made in order to solve the equations at discrete points in space and time and it is this discretisation which leads to the requirement to solve large linear matrix systems. The availability of high speed digital computers together with mathematical research has rendered this problem feasible even for a very large matrix of space points (corresponding to a high resolution in the reservoir description). The formulation and solution of these equations is of importance to the reservoir simulation engineer using the model so that the effects associated with the formulation and numerical solution are understood.

Examples of such numerical effects are saturation oscillations due to the explicit dating of mobility terms or the non-convergence of fully implicit models due to discontinuities in the input data.

3.1.5. *Models as Comparative Tools*

Reservoir Engineers can use numerical reservoir simulation as a tool to improve their insight into the behaviour of the fluids in the reservoir and therefore lead to better reservoir management. However, any reservoir model does not simulate the processes of hydrocarbon production from a reservoir but represents an approximate solution to a mathematical simplification of the very complex reality.

Much of the data, particularly regarding the description of the reservoir away from well control points, is the result of geophysical, geological and engineering interpretation which can lead to the basic model, as initially conceived, giving rise to significant errors in simulated well or field performance. Thus, at an early stage of development, models may be used to quantify the sensitivity of forecasts to some poorly determined parameter by making several predictions each with a different value and, at a later stage when production data becomes available, the model may be adjusted in order to better represent the observed trends.

This refined reservoir description may then be used for optimising hydrocarbon production.

In either case, the model is used in a comparative way, either comparing model with model to investigate sensitivity, or comparing model with field results in order to improve the reservoir description so that meaningful predictions may be made.

3.2.   Equations and Terminology

The individual basic "laws" governing the flow of fluids in a porous medium give rise to simple equations. When these equations are combined and coupled with a simple description of the hydrocarbon phase behaviour together with the effects of immiscible displacement, the result is three coupled nonlinear second order partial differential equations for which, in general, no analytical solutions exist.

In order to reach an approximate solution, the technique of finite differences is (almost exclusively) used and this gives rise to a very large set of linearised equations which must be solved.

The formulation of the finite difference scheme and the solution of the linearised matrix equation using digital computers may produce numerical effects which have important consequences for the practical engineering results of the model.

Awareness of the origin of effects such as numerical oscillation, dispersion of phases or components and the relationship between input data and model performance can significantly change the approach to and the results from a simulation study.

For this reason, an understanding of the techniques used and their potential for affecting the simulation is of some importance to the engineer when setting up and running a reservoir model.

3.2.1.   *Mass Conservation*

The law of mass conservation can be applied to a small volume element with fluid flowing through it along the direction of the x-axis. This is illustrated in Fig. 3.2.1.

MASS BALANCE

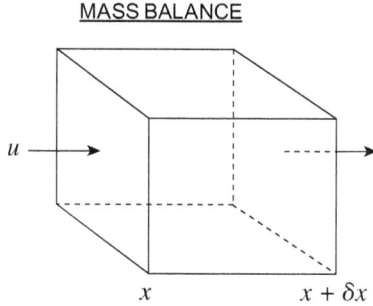

$$\rho_{x+\delta x} u_{x+\delta x} \delta y \delta z \delta t - \rho_x u_x \delta y \delta z \delta t$$
$$= \phi \delta x \delta y \delta z (\rho_{t+\delta t} - \rho_t)$$

$$\frac{\partial(\rho u)}{\partial x} = \frac{\partial(\rho \phi)}{\partial t}$$

Figure 3.2.1. Mass balance.

Applying mass balance at the input and output faces gives

mass in $-$ mass out $=$ accumulation

$$\rho_x u_x \delta y \delta z \delta t - \rho_{x+\delta x} u_{x+\delta x} \delta y \delta z \delta t$$
$$= \phi \delta x \delta y \delta z (\rho_{t+\delta t} - \rho_t). \qquad (3.2.1)$$

Cancelling terms and taking limits $\partial x \to 0$ and $\partial t \to 0$ gives

$$\frac{\partial(\rho u)}{\partial x} = \frac{\partial(\rho \phi)}{\partial t}. \qquad (3.2.2)$$

This equation expresses the conservation of mass for a fluid moving in a 1D system.

3.2.2. *Darcy's Law*

In the middle of the last century, Darcy performed a set of experiments in order to calculate the size of sand filtration beds for the water supply of Dijon. A schematic diagram of his apparatus is shown in Fig. 3.2.2.

He discovered that the rate of flow of water through his equipment was linearly related to the difference in manometer height

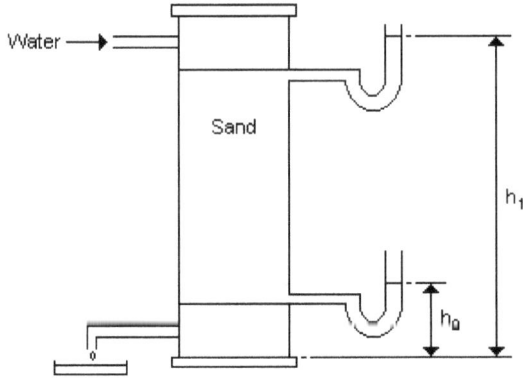

$$u = \frac{k}{\mu}\frac{\partial}{\partial l}(-p + \rho gz)$$

Figure 3.2.2. Darcy's law.

between the inlet and outlet.

$$u_w \propto (h_1 - h_2). \tag{3.2.3}$$

This result was generalised for other experimental configurations and fluids, yielding Darcy's law

$$u = \frac{k}{\mu}\frac{\partial}{\partial l}(p + \rho gz). \tag{3.2.4}$$

It is worth noting that this equation violates Newton's Laws of Motion, as a particle of fluid in a constant potential gradient moves with a constant velocity.

This is similar to the problem of the terminal velocity of a block moving down an inclined plane where gravitational forces are dynamically balanced by frictional forces (see Fig. 3.2.3).

Darcy's law implicitly contains a dynamic balance between the measured potential acceleration force and complex retarding forces. These retarding forces actually depend on the nature of the fluid flow and because of this deviations from this law occur at high flow rates where turbulence effects become important.

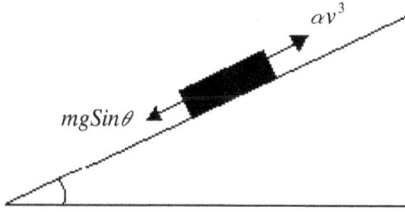

Figure 3.2.3. Dynamic equilibrium.

3.2.3. *Diffusivity Equation*

Generalising the mass conservation and transport equations to 3D gives

$$\nabla(\rho u) = \frac{\partial}{\partial t}(\phi\rho) \tag{3.2.5}$$

and

$$u = \frac{k}{\mu}\nabla(p + \rho g z). \tag{3.2.6}$$

Removing the gravitational term for the moment and combining these equations gives

$$\nabla\frac{\rho k}{\mu}\nabla p = \frac{\partial}{\partial t}(\phi\rho). \tag{3.2.7}$$

Assuming constant porosity and small pressure gradients in a low compressibility fluid, that is

$$\frac{\partial\rho}{\partial t} = \frac{\partial\rho}{\partial p}\frac{\partial p}{\partial t} = \rho c\frac{\partial p}{\partial t}. \tag{3.2.8}$$

Then

$$\nabla^2 p = \frac{\phi\mu c}{k}\frac{\partial p}{\partial t}. \tag{3.2.9}$$

This linearisation gives the diffusivity equation with the diffusivity constant

$$c_{\text{diff}} = \frac{k}{\phi\mu c}. \tag{3.2.10}$$

Note that using cylindrical coordinates to represent flow towards a well gives

$$\frac{1}{r}\frac{\partial}{\partial r}\left[r\frac{\partial p}{\partial r}\right] = \frac{\phi\mu c}{k}\frac{\partial p}{\partial t}. \tag{3.2.11}$$

Which is the radial form of the diffusivity equation, the solution of which is used for analysis of well test data and some analytical aquifers.

In fact, for a black oil system (oil and gas being the two hydrocarbon components) we have the following six equations:

$$\nabla\frac{kk_o}{\mu_oB_o}\nabla(p_o - \rho_ogz) - q_o = \frac{\partial}{\partial t}(\rho_o\phi S_o), \tag{3.2.12}$$

$$\nabla\frac{kk_g}{\mu_gB_g}\nabla(p_g - \rho_gg_z) + \nabla\frac{kk_o}{\mu_oB_o}R_s(p)\nabla(p_o - \rho_ogz) - q_g$$

$$= \frac{\partial}{\partial t}\{(R_s\rho_oS_o + \rho_gS_g)\phi\}, \tag{3.2.13}$$

$$\nabla\frac{kk_w}{\mu_wB_w}\nabla(p_w - \rho_wgz) - q_w = \frac{\partial}{\partial t}(\phi\rho_wS_w), \tag{3.2.14}$$

$$So + S_w + S_g = 1.0, \tag{3.2.15}$$

$$p_o = p_{\text{cow}}(S_w) + p_w, \tag{3.2.16}$$

$$p_o = -p_{\text{cog}}(S_g) + p_g. \tag{3.2.17}$$

These six equations and six unknowns represent the model for fluid flow in the reservoir but, due to coupling (R_s depends on p, p_w depends on p_o as well as $p_{\text{cow}}(Sw)$ and the nonlinearity, (k_o depends on S_o, B_o depends on p_o) in general no analytical solution is possible.

However an approximate solution can be obtained for these equations at preselected space points and for discrete time values by using the method of finite differences.

3.2.4. *Finite Difference*

The method of finite differences makes use of an approximate form for the derivative in order to evaluate the difference equation for each selected space point. For simplicity, consider a function $f(u)$, shown in Fig. 3.2.4, on a uniform grid $u0, u1, \ldots, un-1, un$ where

$$u_{i+1} - u_i = \delta_u. \tag{3.2.18}$$

Using Taylor's theorem we can write

$$f(u_{i+1}) = f(u_i) + \partial u f_i' + \frac{\partial u^2}{2!} f_i'' + \frac{\partial u^3}{3!} f_i''' + \cdots. \tag{3.2.19}$$

$$f(u_{i-1}) = f(u_i) - \partial u f_i' + \frac{\partial u^2}{2!} f_i'' - \frac{\partial u^3}{3!} f_i''' + \cdots. \tag{3.2.20}$$

Using this to approximate derivatives we obtain

$$\frac{\partial f^f}{\partial u} = \frac{f(u_{i+1}) - f(u_i)}{\delta u} + O(\delta u), \tag{3.2.21}$$

$$\frac{\partial f^b}{\partial u} = \frac{f(u_i) - f(u_{i-1})}{\delta u} + O(\delta u), \tag{3.2.22}$$

$$\frac{\partial^2 f}{\partial u^2} = \frac{f(u_{i-1}) + f(u_{i+1}) - 2f(u_i)}{\delta u^2} + O((\delta u)^2), \tag{3.2.23}$$

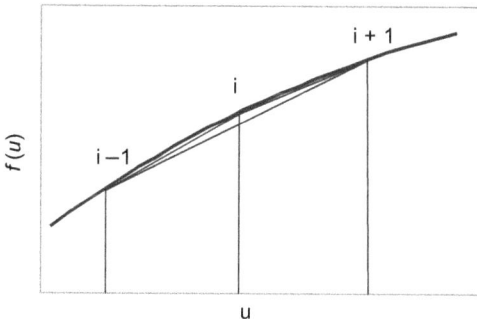

$$\frac{\partial f^f}{\partial u} = \frac{f(u_{i+1}) - f(u_i)}{\delta u} + o(\delta u)$$

Figure 3.2.4. Finite difference.

where the superscripts b and f refer to backwards and forwards as the directions for computing the gradients.

If we apply these approximations to a simplified oil equation in one space dimension, then we obtain

$$\frac{1}{\delta x^2} \left\{ \left[\frac{kk_o}{\mu_o B_o} \right]_{i+1/2} (p_{i+1} - p_i) - \left[\frac{kk_o}{\mu_o B_o} \right]_{i-1/2} (p_i - p_{i-1}) \right\}$$

$$-\bar{\bar{q}}_o = \frac{1}{\delta t} \left\{ \left[\frac{\phi S_o}{B_o} \right]_i^{n+1} - \left[\frac{\phi S_o}{B_o} \right]_i^{n} \right\}, \qquad (3.2.24)$$

where for each space point we have a similar difference equation for each of the three fluid components.

Multiplying through by the volume of the element converts the specific rate term to mass per unit time and gives:

$$\frac{\delta y \delta z}{\delta x} \left\{ \left[\frac{kk_o}{\mu_o B_o} \right]_{i+1/2} (p_{i+1} - p_i) - \left[\frac{kk_o}{\mu_o B_o} \right]_{i-1/2} (p_i - p_{i-1}) \right\}$$

$$-q_o = \frac{V}{\delta t} \left\{ \left[\frac{\phi S_o}{B_o} \right]_i^{n+1} - \left[\frac{\phi S_o}{B_o} \right]_i^{n} \right\}. \qquad (3.2.25)$$

The superscript n refers to the start of the time step and $n + 1$ to the end of the time step.

Thus because of the errors inherent in the approximation of the derivatives, for each space point and in each component, errors are introduced into both the space and time parts of the equations.

These errors are called truncation errors.

The space truncation errors being of order δx^2 are usually small provided that the grid of space points is selected to be reasonably regular. For adjacent grid cells with large differences in spatial extent, space truncation errors can become important.

More often, time truncation errors are more important as the derivative approximation is correct only to order δt and the error grows with the time step size.

3.2.5. *Implicit and Explicit Formulation*

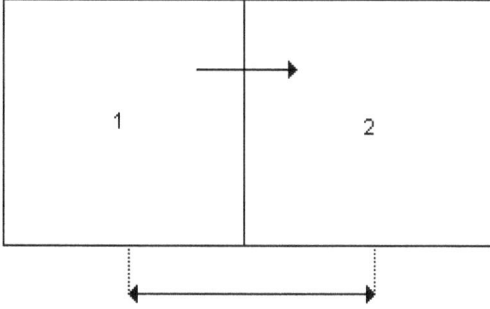

Figure 3.2.5. Flow between two cells.

EXPLICIT FORMULATION

$$\delta\rho = \frac{Ak}{\mu_o B_o}\frac{\Delta p}{\Delta x}\delta t\frac{1}{A\Delta x_\phi},$$

$$\delta t \prec \frac{\Delta x^2}{2}\frac{\phi\mu_o c}{k}. \qquad (3.2.26)$$

We have written for the oil equation

$$\frac{\delta y \delta z}{\delta x}\left\{\left[\frac{kk_o}{\mu_o B_o}\right]_{i+1/2}(p_{i+1} - p_i) - \left[\frac{kk_o}{\mu_o B_o}\right]_{i-1/2}(p_i - p_{i-1})\right\}$$

$$-q_o = \frac{V}{\delta t}\left\{\left[\frac{\phi S_o}{B_o}\right]_i^{n+1} - \left[\frac{\phi S_o}{B_o}\right]_i^n\right\}. \qquad (3.2.27)$$

This difference equation is formulated to obey Darcy's law and to conserve mass. The computed flux between two space points will depend on the dating of the mobility term and the pressure difference term.

If we consider two cells completely filled with oil with an initial pressure difference Δp (shown schematically in Fig. 3.2.5), then from

Darcy's law we can compute the instantaneous flow between cells

$$N = \frac{Ak}{\mu_o Bo} \frac{\Delta p}{\Delta x} \delta t \qquad (3.2.28)$$

and the resulting change in density is

$$\delta p = \frac{Ak}{\mu_o Bo} \frac{\Delta p}{\Delta x} \delta t \frac{1}{A\Delta x\phi}. \qquad (3.2.29)$$

For a total compressibility of c, then the change in pressure in the cell as a result of fluid flow is:

$$\delta p = \frac{k}{\phi \mu_o c} \frac{\Delta p}{\Delta x^2} \delta t. \qquad (3.2.30)$$

For stability, δp must be less than $\Delta p/2$
    or

$$\delta t < \frac{\Delta x^2}{2} \frac{\phi \mu_o c}{k}. \qquad (3.2.31)$$

Using:

$$\Delta x = 10^4 \text{cm}$$

$$\mu_o = 1\text{cp}$$

$$c = 10^{-4} \text{atm}^{-1}$$

$$k = 1\text{darcy}$$

$$\Phi = 0.2$$

then $\delta t < 10^3$ s or approximately 15 min.

If this stable time step is exceeded, then oscillations (possibly amplified) will occur.

In order to relax the stability criterion to provide useful time step size, the pressure must be treated implicitly. The difference between the explicit and implicit flux calculation is illustrated in Fig. 3.2.6.

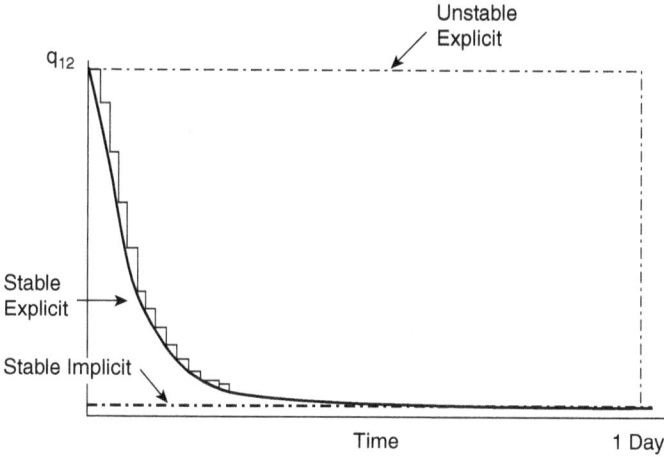

Figure 3.2.6. Implicit and explicit.

Writing the pressure difference at the new time level, we have

$$
\frac{\partial y \partial z}{\delta x} \left\{ \left[\frac{k k_o}{\mu_o B_o} \right]_{i+1/2} (p_{i+1} - p_i)^{n+1} \right.
$$

$$
\left. - \left[\frac{k k_o}{\mu_o B_o} \right]_{i-1/2} (p_i - p_{i-1})^{n+1} \right\} - q_o
$$

$$
= \frac{V}{\delta t} \left\{ \left[\frac{\phi S_o}{B_o} \right]_i^{n+1} - \left[\frac{\phi S_o}{B_o} \right]_i^{n} \right\}, \tag{3.2.32}
$$

Linearising the changes in Φ and $1/B_o$ with respect to pressure and writing

$$
h_i = -\delta t \delta y \delta z \left\{ \left[\frac{k}{\mu_o B_o} \right]_{i+1/2}^{n} (p_{i+1}^n - p_i^n) \right.
$$

$$
\left. - \left[\frac{k}{\mu_o B_o} \right]_{i-1/2}^{n} (p_i^n - p_{i-1}^n) \right\}, \tag{3.2.33}
$$

$$
a_{i-} = \left[\frac{k}{\mu_o B_o} \right]_{i-1/2} \delta t \delta y \delta z, \tag{3.2.34}
$$

$$a_{i+} = \left[\frac{k}{\mu_o B_o}\right]_{i+1/2} \delta t \delta y \delta z, \qquad (3.2.35)$$

$$a_i = a_- + a_+ + \left(\frac{1}{B_o}\frac{\partial \phi}{\partial p} - \phi \frac{\partial \left(\frac{1}{B_o}\right)}{\partial p}\right) V \delta x. \qquad (3.2.36)$$

Then we can write

$$b_i = a_{i-}\Delta p_{i-1} + u_i \Delta p_i + u_{i+}\Delta p_{i+1}, \qquad (3.2.37)$$

where the change in pressure over the time step is

$$\Delta p_i = p_i^{n+1} - p_i^n. \qquad (3.2.38)$$

For a 1D problem with m space points we have m such equations

$$\begin{bmatrix} a_1 & a_{1+} & \cdot & \cdot & \cdot & \cdot \\ a_{2-} & a_2 & a_{2+} & \cdot & \cdot & \cdot \\ \cdot & a_{3-} & a_3 & a_{3+} & \cdot & \cdot \\ \cdot & \cdot & \cdot & \cdot & \cdot & \cdot \\ \cdot & \cdot & \cdot & \cdot & \cdot & \cdot \\ \cdot & \cdot & \cdot & \cdot & a_{m-} & a_m \end{bmatrix} \begin{bmatrix} \Delta p_1 \\ \Delta p_2 \\ \Delta p_3 \\ \cdot \\ \cdot \\ \Delta p_m \end{bmatrix} = \begin{bmatrix} b_1 \\ b_2 \\ b_3 \\ \cdot \\ \cdot \\ b_m \end{bmatrix} \qquad (3.2.39)$$

or in a more compact form

$$A\Delta p = b, \qquad (3.2.40)$$

where A is the coefficient matrix and Δp and b are vectors of length m.

The value of the pressure change for any individual block can no longer be computed in isolation but its value is implicitly included within the solution of the complete set of equations.

This leaves open the question of the dating of the mobility terms. Depending upon the simulation problem, this might be stable using schemes explicit in saturation.

Fully implicit schemes evaluate all coefficients on the left-hand side of the equations at the new time level $n + 1$.

IMPES schemes (implicit pressure, explicit saturation) evaluate saturation dependent terms at the start of the time step. There are many schemes which lie between these in terms of the degree of implicitness of the formulation.

Explicit schemes are quick to evaluate per time step and require little computer storage space but have stability constraints which may lead to oscillations or divergence of the solution.

Implicit schemes are slower per time step but are unconditionally stable and for this reason have become the norm for black oil models.

3.2.6. *Dispersion and Weighting*

The mobility term in the difference equation refers to the mobility of the fluids at the interface between cells. In order to compute this we would need the saturation and pressure at the cell interface which by definition is not one of the calculation nodes. The usual approach is to assign the mobility at the face to be that computed at the cell centre of the upstream (higher potential) block.

This leads to fluid within the upstream finite difference cell being dispersed and can result in poor simulation of sharp saturation (or component) fronts (see Fig. 3.2.7). The severity of the numerical dispersion which is produced depends on the size of the finite difference mesh chosen.

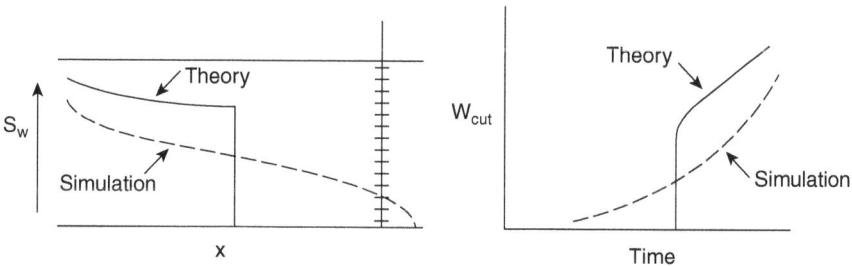

Figure 3.2.7. Dispersion.

3.2.7. *Nonlinearity and Outer Iterations*

In order to set up the discretised equation for fluid flow

$$A\Delta p = b \tag{3.2.41}$$

assumptions were made regarding the linear dependence of porosity and the inverse of the volume factor with respect to pressure.

A consequence of this discretisation is that the value of $\boldsymbol{B}_o^{n+1}$ is required.

The technique used to compute this is a Newton–Raphson approach where

$$\frac{1}{\boldsymbol{B}_o^{k+1}} = \frac{1}{\boldsymbol{B}(p^k)} + \left[\frac{\partial 1/\boldsymbol{B}_o}{\partial p}\right]^k \partial p^{k+1}. \tag{3.2.42}$$

The superscript k refers to the iteration level and δp^{k+1} is the pressure change over the iteration.

The estimate of the value at the end of the iteration is expressed as a linear function of the change in pressure over that iteration.

The first iteration uses the start of time step values for the initial start point and gradient.

At the end of the first iteration, $1/B_o$ and its derivative are re-evaluated at the new pressure value. This iteration procedure

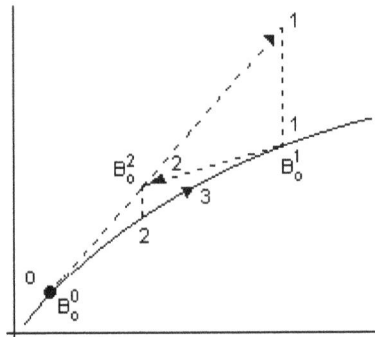

$$\frac{1}{B_o^{k+1}} = \frac{1}{B_o(p^k)} + \left(\frac{\partial 1/B_o}{\partial p}\right)^k \delta p^{k+1}$$

Figure 3.2.8. Newton iteration.

stops when

- all the values of δp are smaller than some tolerance, and
- the material balance is within a tolerance, or
- the maximum number of iterations has been reached.

The iterative procedure is illustrated in Fig. 3.2.8.

If, for some cells, the value of the formation volume factor at the new time level, $\boldsymbol{B}_o^{n+1}$, is not convergent, then the densities used in the difference equation will be incompatible and the solution will not conserve mass. This will result in a material balance error.

3.2.8. *Linear Solvers*

Having written the equation

$$A\Delta p = b, \qquad\qquad (3.2.43)$$

the solution must be obtained for the vector of changes to the pressure. For reservoir simulation, special purpose linear solvers have been formulated in order to maximise the efficiency of this task.

This effort has been necessary because of the size of the problem which for a fully implicit black oil model with one thousand cells would contain nine million matrix elements if the A matrix was treated as a general square matrix.

The solution algorithms developed take account of not only the sparseness of the matrix but also of the regular nature of the occurrence of matrix elements.

The actual location of the matrix elements depends on the indexing scheme adopted for the finite difference space points and this in turn can affect the number of arithmetic operations needed to invert the matrix.

Some indexing schemes are shown in Figs. 3.2.9 and 3.2.10.

If the A matrix can be written as

$$A = LU, \qquad\qquad (3.2.44)$$

where L is lower triangular and U is upper triangular, then the problem may be written

$$LU\Delta p = b \qquad\qquad (3.2.45)$$

1	2	3	4	5
6	7	8	9	10
11	12	13	14	15

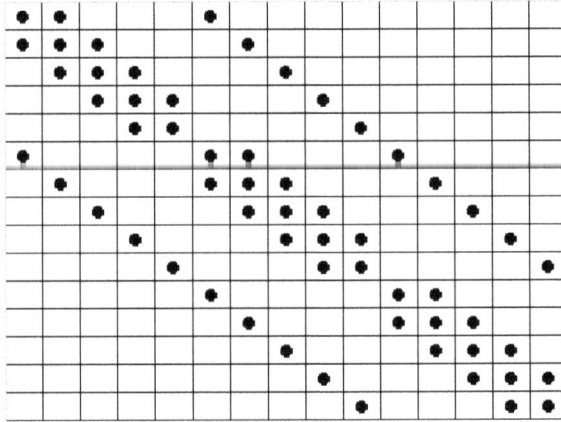

Figure 3.2.9. Matrix indexing scheme 1.

1	4	7	10	13
2	5	8	11	14
3	6	9	12	15

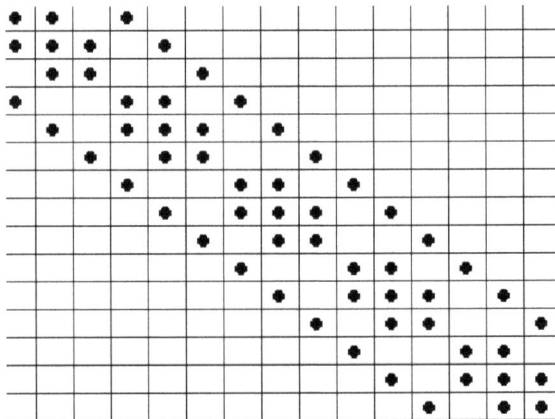

Figure 3.2.10. Matrix indexing scheme 2.

or

$$U\Delta p = L^{-1}b \tag{3.2.46}$$

and hence the back substitution

$$\Delta p = U^{-1}L^{-1}b. \tag{3.2.47}$$

This corresponds to performing a forward elimination and a back substitution giving the vector of pressure changes precisely.

The process of elimination involves manipulation of matrix rows and requires a number of arithmetic operations dependent on the difference in index between the outermost matrix elements in each row. This is referred to as the "matrix band width" and is related to the ordering scheme used and the number of cells in each dimension.

For a model with n_x, n_y and n_z cells in the x, y and z direction respectively, ordering first in the x direction gives a band width w

$$w = 2n_y n_z + 1. \tag{3.2.48}$$

The classical approach of forward elimination and back substitution, also known as Gaussian elimination, is called a direct solution. The number of arithmetic operations involved in such a solution depends on the band width and is approximated as

$$N \propto n_x n_y^3 n_z^3. \tag{3.2.49}$$

A different approach is to attempt an approximate LU decomposition by iterative means. The number of arithmetic operations depends on the formulation of the iterative technique but can be approximated as

$$N \propto n_{\text{iter}} n_x n_y n_z, \tag{3.2.50}$$

where n_{iter} is the number of iterations required to converge to the linear solution.

Although it is beyond the scope of this course to treat the many iterative schemes, the fundamentals may be illustrated by considering one such scheme.

For a 2D problem, we can write

$$A\Delta p = b \tag{3.2.51}$$

as

$$\begin{bmatrix} T_1 & H_1 & \cdot & \cdot \\ G_2 & T_2 & H_2 & \cdot \\ \cdot & G_3 & T_3 & H_3 \\ \cdot & \cdot & G_4 & T_4 \end{bmatrix} \begin{pmatrix} x_1 \\ x_2 \\ x_3 \\ x_4 \end{pmatrix} = \begin{pmatrix} B_1 \\ B_2 \\ B_3 \\ D_4 \end{pmatrix}, \tag{3.2.52}$$

where T_i is a tri-diagonal matrix of dimension ℓ, where ℓ is the length of the model. H_i and G_i are diagonal matrices of dimension ℓ and x_i and B_i are vectors of length ℓ of the change in pressure and b respectively.

One method of resolving this set of equations is to make a direct solution of the tri-diagonal submatrices as follows

$$x_1^{m+1} = T_1^{-1}(B_1 - H_1 x_2^m), \tag{3.2.53}$$

$$x_2^{m+1} = T_2^{-1}(B_2 - G_2 x_1^{m+1} - H_2 x_3^m), \tag{3.2.54}$$

$$x_3^{m+1} = T_3^{-1}(B_3 - G_3 x_2^{m+1} - H_3 x_4^m), \tag{3.2.55}$$

$$x_4^{m+1} = T_4^{-1}(B_4 - G_4 x_3^{m+1}). \tag{3.2.56}$$

Here m is the index of the iteration level for the linear solver.

This iterative procedure is repeated until all the elements of the vector are smaller than some tolerance.

$$(x^{m+1} - x^m). \tag{3.2.57}$$

The method illustrated here uses the latest up dated values of the solutions of x in order to modify the right-hand side and is known as a line Gauss–Seidel procedure.

This iterative procedure is at the level of the solution of the linearised equations and consequently the iterations, index m, are referred to as "inner iterations" compared to the "outer iterations", index k, required to resolve the nonlinearities.

The overall iterative scheme is shown in Fig. 3.2.11 where the relationship between inner and outer iterations can be seen.

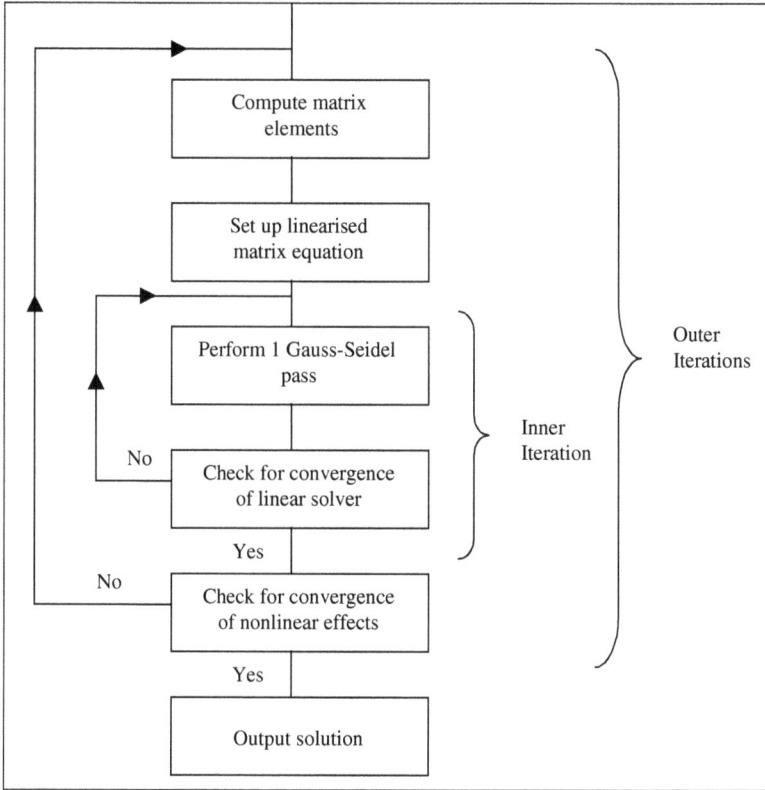

Figure 3.2.11. Overall iteration scheme.

3.3. Buckley–Leverett Displacement

If we consider the flow of an incompressible fluid (water say) displacing a different, immiscible fluid from a 1D porous system at constant fluid rate, then applying conservation of mass gives

$$\left[\frac{\partial q_w}{\partial x}\right]_t = -A\phi \left[\frac{\partial S_w}{\partial x}\right]_x. \tag{3.3.1}$$

We can write the differential for the water saturation as

$$dS_w = \left[\frac{\partial S_w}{\partial x}\right]_t dx + \left[\frac{\partial S_w}{\partial x}\right]_x dt. \tag{3.3.2}$$

If we consider the movement of a particular water saturation contour through the system then for that contour $dS_w = 0$. Then

$$\left[\frac{\partial S_w}{\partial t}\right]_x = -\left[\frac{\partial S_w}{\partial x}\right]\left(\frac{dx}{dt}\right)_{S_w}. \tag{3.3.3}$$

Defining the fractional flow of water as

$$f_w = \frac{q_w}{q_{\text{total}}}, \tag{3.3.4}$$

which for water–oil displacement is

$$f_w = \frac{q_w}{q_w + q_o}, \tag{3.3.5}$$

then, because the total flow rate is constant,

$$\frac{\partial q_w}{\partial x} = q_{\text{total}}\frac{\partial f_w}{\partial x} \tag{3.3.6}$$

and

$$\frac{\partial f_w}{\partial x} = \frac{\partial f_w}{\partial S_x}\frac{\partial S_w}{\partial x}. \tag{3.3.7}$$

Substituting Eqs. (3.3.3) and (3.3.7) into Eq. (3.3.1), we obtain

$$q_{\text{total}}\frac{\partial f_w}{\partial S_w}\frac{\partial S_w}{\partial x} = A\phi\frac{\partial S_w}{\partial x}\left[\frac{dx}{dt}\right]_{S_w}. \tag{3.3.8}$$

Cancelling and rearranging gives

$$v_{S_w} = \left[\frac{dx}{dt}\right]_{S_w} = \frac{q_{\text{total}}}{A\phi}\left[\frac{df_w}{dS_w}\right]_{S_w}. \tag{3.3.9}$$

That is, the velocity of the contour is proportional to the gradient of the fraction flow at that water saturation.

The water fractional flow for a typical North Sea oil–water system is shown in Figs. 3.3.1 and 3.3.2. This function results in a mathematical inconsistency because the velocity of the high water saturation is greater than the velocity of the leading lower saturation contours.

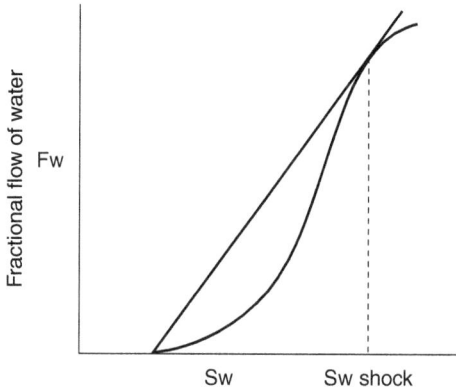

Figure 3.3.1. Fractional flow.

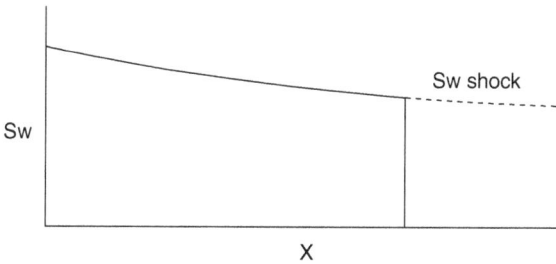

Figure 3.3.2. Buckly Leverett displacement.

This results in the formation of a shock front or piston of water moving through the 1D system and leads to an efficient displacement of the oil.

This analytical treatment can be used to compare with the results of numerical simulation in order to investigate the effects of numerical dispersion in simple systems.

It should be recognised that although in displacement of water by oil in oil fields there are dispersive contributions from capillary pressure, vertical and lateral inhomogeneities of porosity and permeability as well as the effect of gravity, these dispersive terms do not behave in the same manner as the numerical dispersion.

3.4. Reservoir Models

Although all reservoirs are different and so the models built to study them are different, the components of the basic input data and the category of model are similar between all reservoir simulators.

3.4.1. *Model Components*

All reservoir models have components of data which correspond to the major activities performed during reservoir simulation.

3.4.1.1. *Reservoir description*

The description of the reservoir requires

- the reservoir geometry,
- the rock properties,
- the multi-phase fluid flow properties,
- the fluid-phase behaviour properties,
- the aquifer properties.

3.4.1.2. *Initialisation*

In order to set the initial phase saturations and pressures, the model requires

- the position of fluid contacts,
- the capillary pressure function,
- any spatial variation of fluid properties,
- a reference pressure at a specific depth.

3.4.1.3. *Model control*

Although frequently set by default, models have a large range of control parameters which determine

- the finite difference formulation,
- the linear solution method,
- the number of nonlinear iterations,
- the required accuracy of the solution.

3.4.1.4. *Production data*

This consists of the time dependent data specifying

- well location and type,
- perforation interval and parameters,
- well target rates,
- well or group constraints,
- actions resulting from some violation.

3.4.1.5. *Output*

The user can request different types of reports or data to be output

- time step summary,
- well summary,
- region summary,
- cell by cell values of variables,
- graphics files,
- pseudo files,
- restart files.

The detailed form of the input data depends on the type and input format specification of the model which is being used. As the complexity of the physical processes treated by the model increases so does the quantity of data require to define the processes. Although most simulators provide some sort of data screening, they are by no means infallible and the requirement for consistency in the input data is the responsibility of the engineer setting up the model.

3.4.2. *Model Types*

The approach to reservoir modelling is one by which specific behaviour of the field (or part of the field) may be studied in some detail and simplified in some way so that the same behaviour can be accurately preserved in a model which would not, *a priori*, have sufficient resolution to simulate the effect. However, this lower resolution model can include many wells and field constraints which it would not have been feasible to be included in the initial high resolution model.

The principal types of model used are described below.

3.4.2.1.   *Cross-sectional models*

One of the major controlling influences on the production from oil and gas reservoirs is the distribution of the permeability through the reservoir section and the degree of vertical communication between the reservoir layers.

The presence of a thin but continuous layer of very high or low permeability may have an effect on the production characteristics of the field (or sector) which must be faithfully reproduced in a model where the layer thickness could be many times the magnitude of the particular zone.

A cross-sectional model permits the production characteristics to be investigated and for the engineer to represent the resulting reservoir behaviour in a form suitable for inclusion in the coarse model.

In certain fields, a representative cross-section may be used to investigate future production and the results scaled in some way to give the expected field behaviour.

3.4.2.2.   *Sector models*

The use of a representative cross-section whose results are scaled to give the behaviour of the field, is actually a special case of the sector model where a representative microcosm is studied in some detail and the field profile is computed by integrating the results in some way, including the effects of scheduling and constraints.

Models of symmetry elements of pattern floods and isolated fault block models are examples of such sectors.

Sector models are usually run when the flux across the external boundary of the model is zero or negligible. In certain cases it is possible to define the fluxes in some way and perform a sector model study with externally imposed flux values at each time step. Such models have been used to study the effects of several oil fields producing from a zone connected by a common aquifer.

3.4.2.3. *Cylindrical single well models*

Single well models are frequently used to improve the reservoir description as a result of some effect in the pressure behaviour of a well test. They are also used for investigating the dependence of the produced fluids on rate and contact position either for optimisation of the completion or for including the results in a coarse simulation model.

3.4.2.4. *Full field models*

The individual high resolution models are used for detailed examination of a particular zone or effect, but the overall interaction of production and injection in all other wells as well as the imposition of group or field constraints is studied by using a full field model. The high resolution model behaviour is reflected in the full field model by the use of up-scaling to represent the fluid flow or production characteristics of a well.

Despite the use of up-scaling to integrate the detailed reservoir behaviour into a form for full field simulation, it is usual for the coarse model to have many tens of thousands of active cells and models with hundreds of thousands of grid blocks are quite normal.

Producing coherent input data and analysing the results of models of this size is a major task.

3.5. Grid Systems

In Sec. 3.2.4, we wrote the finite difference form of the 1D, equally spaced, Cartesian diffusivity equation for oil. Simplifying the notation, we can write this as

$$T_{x_{i+}}(p_{i+1} - p_i) - T_{x_{i-}}(p_i - p_{i-1}) = \frac{1}{\delta t}\Delta_t \left[\frac{\phi S_o}{B_o}\right], \qquad (3.5.1)$$

where T is the transmissibility term and Δ_t is the time difference operator. In cylindrical coordinates, the linearised diffusivity equation becomes

$$\frac{1}{r}\frac{\partial}{\partial r}\left[r\frac{\partial p}{\partial r}\right] = \frac{\phi\mu c}{k}\frac{\partial p}{\partial t}. \tag{3.5.2}$$

This may be discretised in the same way as the Cartesian equation, giving

$$\frac{1}{r_i\delta r_i}\left\{\left[\frac{kk_o}{\mu_o B_o}\right]_{i+1/2}r_{i+1/2}(p_{i+1}-p_i)\right.$$

$$\left.-\left[\frac{kk_o}{\mu_o B_o}\right]_{i-1/2}r_{i-1/2}(p_i-p_{i-1})\right\}$$

$$= \frac{V}{\delta t}\left\{\left[\frac{\phi S_o}{B_o}\right]_i^{n+1}-\left[\frac{\phi S_o}{B_o}\right]_n^i\right\}, \tag{3.5.3}$$

where $r_{i-1/2}$ represents a point between point i and $i-1$ where the mobility term is computed. It is not clear that for radial coordinate systems the arithmetic average location is a good choice. For steady state inflow into a well it can be shown that the logarithmic mean

$$r_{i+1/2} = \frac{r_{i+1}-r_i}{in(r_{i+1}/r_i)} \tag{3.5.4}$$

is a better choice.

For some choice of transmissibility centre, we can write

$$T_{r_{i+}}(p_{i+1}-p_i)-T_{r_{i-}}(p_i-p_{i-1}) = \frac{1}{\delta t}\Delta_t\left[\frac{\phi s_o}{B_o}\right] \tag{3.5.5}$$

which is identical in form to the Cartesian case, except that the different geometrical configuration has been incorporated into the *a priori* computation of the transmissibility equation.

This property is quite fundamental to most reservoir simulators as (with minor modifications) more complex coordinate systems can be dealt with in a standard simulator by a preliminary treatment of the geometrical terms which affect the transmissibility calculation.

Note that whilst a constantly spaced 1D Cartesian system and a logarithmically spaced 1D radial system are completely analogous, a 2D areal Cartesian model and a 2D areal radial model are not.

This is due to the cyclic nature of the boundary conditions for cylindrical models, discretised in the azimuthal coordinate, because of the connection between the first and the last sectors.

3.5.1. *Cartesian Systems*

Cartesian models (see Fig. 3.5.1), built on rectangular grid cells, are one of the most common type of models. This is because they are easy to set up and the results are easily interpreted. The potential problems with these models are that the choice of gridding may not be optimum and that grid orientation effects may be present.

Never the less, the ease of setting up the model, together with the ease of relating the results to field behaviour are the main reasons for these models being the first choice for many simulation projects.

The choice of the dimensions of the grid is a compromise between accuracy and cost (or computer execution time). The size of a grid for any particular model depends on:

(1) the definition required for the structural description,
(2) the location of fluid contacts,
(3) the proposed well density.

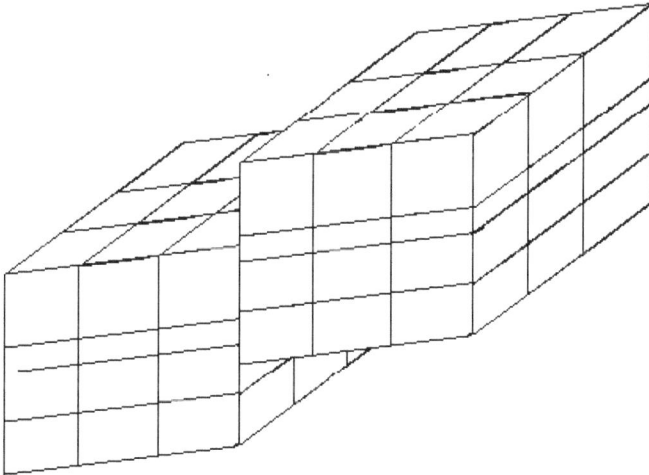

Figure 3.5.1. 3D Cartesion system.

Clearly the mesh size should also depend on the amount and quality of data available. There is little point in constructing a detailed model with many thousands of grid cells for a prospect with only the discovery well and coarsely spaced 2D seismic data available.

3.5.2.   *Cylindrical Systems*

A schematic cylindrical system is displayed in Fig. 3.5.2. As demonstrated above, the mathematical formulation (apart from cyclic boundary conditions) is independent of the coordinate system chosen. Historically this was not the case, as the numerically more sensitive cylindrical coordinate models were frequently designed to have a higher degree of implicitness which could not feasibly be run on a large scale 3D model.

These single well models (frequently called coning models, because the study of coning phenomena was one of their primary uses) have largely been incorporated into the standard commercially available systems as an option for the coordinate system choice. This is possible because the current generation of black oil reservoir

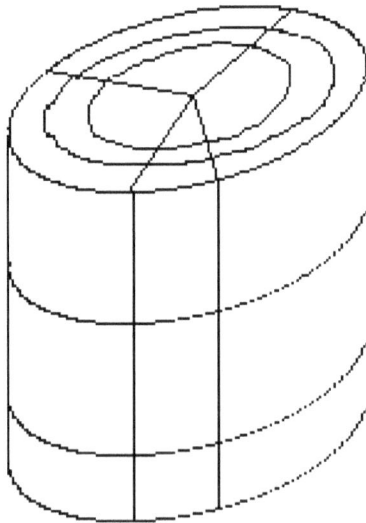

Figure 3.5.2.   3D Radial cylindrical system.

simulators have available the option for a fully implicit formulation of the difference equations.

3.5.3. *Stream Line Grids*

In the same way that mapping the radial system on to the Cartesian system with a one-to-one correspondence between grid cells, permits the simulator to operate independently of the coordinate system, once the pore volume and transmissibility values have been correctly calculated, any system which preserves the cell indexing and inter connection scheme can be used provided that the appropriate geometrical calculations have been made.

One such choice is the curved set of grids obtained from using streamlines and iso-potentials for defining the grids. These grids are however analytically calculable for a restricted set of well configurations and fluid mobility ratios and maybe distorted by reservoir inhomogeneity or changes in field operation. Because of this, a simplified approach may be used approximating the flow lines by straight lines. Such a system is shown in Fig. 3.5.3.

However, in certain circumstances they can provide a cost effective solution to rather complex problems but great care must be taken as the applicability of the results from the simulator depend critically on the accuracy of the choice of stream lines.

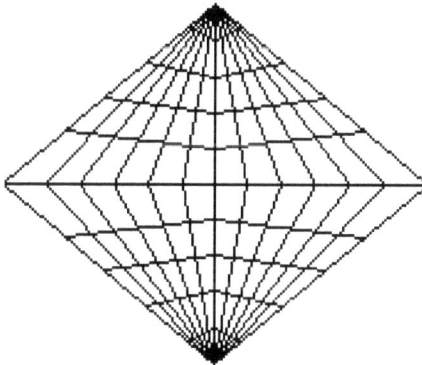

Figure 3.5.3. Approximate stream line system.

3.5.4.   *Special Connections*

The preservation of the mathematical mapping between the different coordinate systems permits the simulator to operate because the matrix coefficients can be computed, using predefined geometrically dependent terms, and the location of the matrix elements is unperturbed.

This is not quite true for the cylindrical system, where the connection between the first and the last azimuthal sector requires some additional matrix elements. These however appear in strictly defined positions and can be accommodated in the efficient solution of the linearised equations.

Further additions to the matrix are necessary for the inclusion of well constraints. If a well, which is perforated over several non-communicating layers, is producing at it maximum permitted rate then the contribution from each layer depends implicitly on the others.

$$q_{\text{total}} = q_{l1} + q_{l2} + \cdots + q_{lk}. \tag{3.5.6}$$

This introduces matrix elements connecting all the perforated cells in the well. The location of these additional matrix elements is clearly less well defined than those produced by cyclic boundary conditions and may also be time dependent as wells are perforated.

Geological pinch out also juxtaposes cells not adjacent in the normal indexing schemes. These effects are shown schematically in Fig. 3.5.4.

The treatment of flow across fluid transmitting geological faults (see Fig. 3.5.5) can also introduce matrix elements which are additional to the usual regular pattern. This is because a fault having a throw equal to half of the thickness of a layer will reduce the transmissibility in that layer but add transmissibility matrix elements corresponding to the additional communications between layers across the fault.

Although the occurrence of these elements is not (usually) time dependent, they may be large in number for a model of a large faulted field, and the disruption to the matrix pattern may be severe.

Cross flow

Pinch out

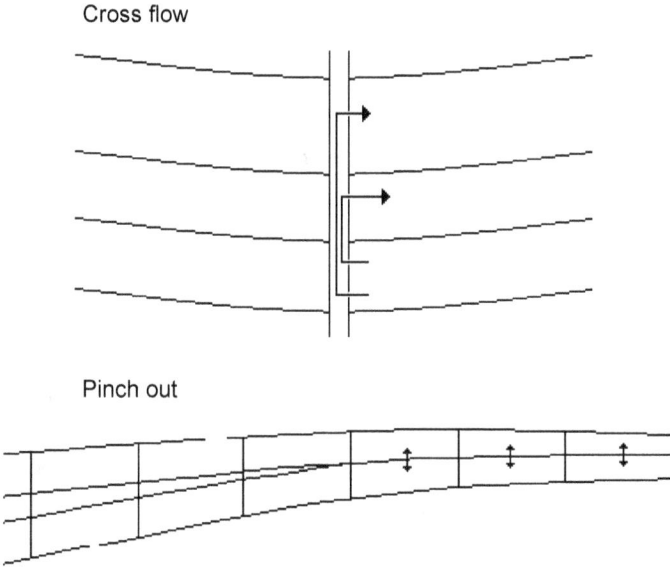

Figure 3.5.4. Well layer flows and pinch out.

Fault

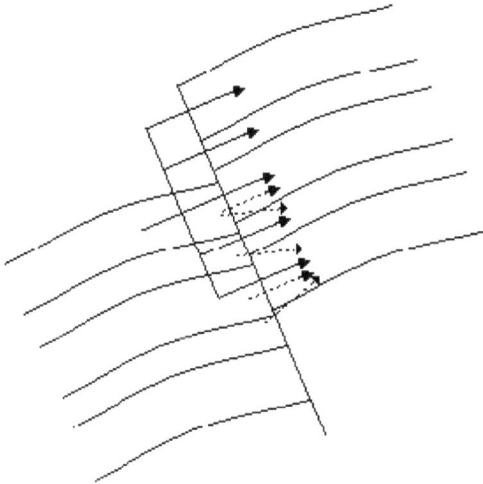

Figure 3.5.5. Flow across faults.

This can have severe consequences for the speed of each inner iteration and also on the convergence properties of the matrix.

3.5.5. *Corner Point Representation*

The extension of the mapping of the areal coordinate system to a distorted grid, following the fault directions, as well as the requirement to model non-vertical faults leads to grid cells which may have a general six faced shape. It is no longer possible to describe the cell geometry simply by its extension in 3D and the solution adopted is to give the simulator the space coordinates of the eight points corresponding to the corners. Note that the neighbouring grid block is also specified using eight corner points and there is no requirement that the points between the cells should be coincident.

Provided that the transmissibilities can be calculated between neighbouring cells, then the distorted grid is treated by the simulator in the same way as any other geometrical mapping.

3.5.6. *Local Grid Refinement*

The usual approach to the spatial discretisation is to choose a set of grid cells defined by lines superimposed on the structure map. These lines may be curved, if a non-cartesian system is being used, but, in order to preserve the matrix structure, they are continuous.

The ability to deal with grid cells having more than the six nearest neighbours was required for the treatment of flow across geological faults and the same technique can be used to set up the matrix corresponding to a locally refined grid. Examples of the use of such grids are shown in Fig. 3.5.6.

Use of local grid refinement automatically adds special connections into the matrix. An example of this is shown in Fig. 3.5.7, where the extra transmissibility values corresponding to both the refined grid and its connection to the global grid may be seen.

Because the matrix structure may be seriously disturbed, the time taken per linear iteration and the number of iterations required for convergence may be increased substantially.

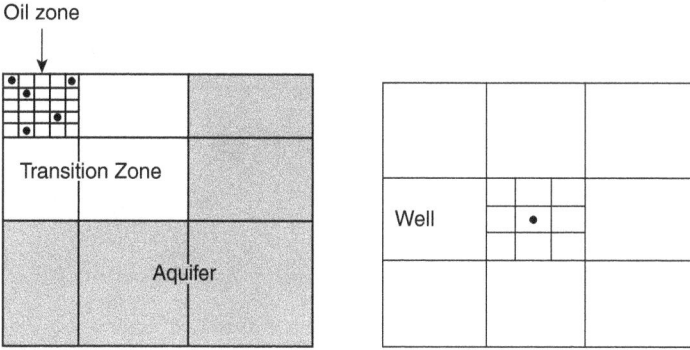

Figure 3.5.6. Local grid refinement.

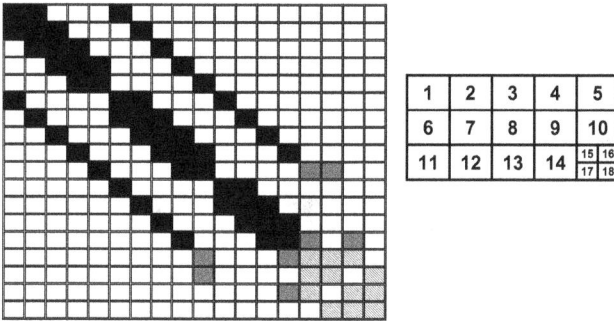

Figure 3.5.7. Local grid refinement showing the effect on the matrix structure.

3.5.7. *Unstructured Grids*

As the complexity of the grid increases, then there is a corresponding loss of simplicity in the matrix structure which when taken to the limit means that the coefficient matrix becomes a generalised sparse matrix with little if any evident structure. Such unstructured grids may result from a gridding technique using perpendicular bisectors (PEBI) of lines joining a set of control points. These control points become the grid node and the geometry associated with the node is determined by the set of PEBI lines surrounding each one. Figure 3.5.8 shows an example of a PEBI grid. Notice that some parts of the grid look almost Cartesian whilst some cells have five or six nearest neighbours.

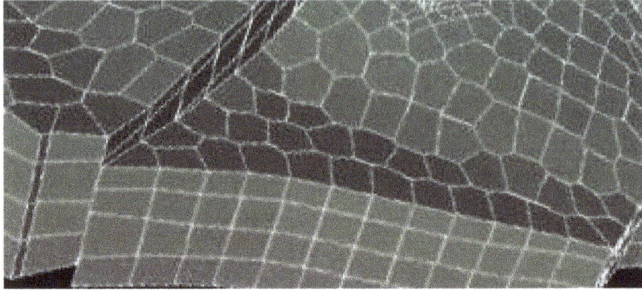

Figure 3.5.8. Example of unstructured pebi grid.

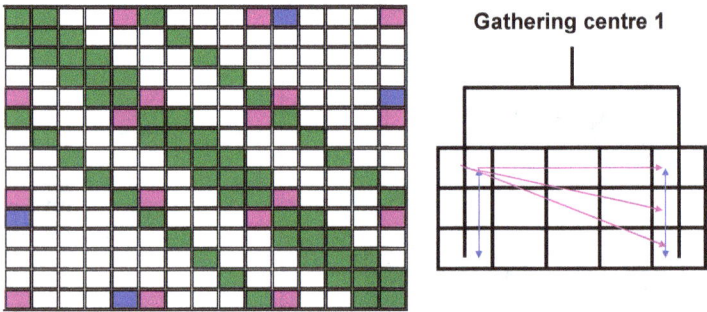

Figure 3.5.9. Additional matrix elements due to surface constraints.

3.5.8. *Surface Constraints*

In the same way that complex gridding complicates the matrix structure, adding surface constraints can add many matrix coefficients not normally present in the set of equations governing inter-block flow. A simple example of this is shown in Fig. 3.5.9, where the constraint at the manifold propagates back into the subsurface matrix connecting many blocks with special connections.

Most present day simulators do not include the gathering centre constraints in the matrix as the special connection overhead is too great so the constraint is wound in through the Newton iteration scheme. However, a new generation of simulators (Landmark's nexus and Schlumberger's Intersect) built to handle unstructured grids can include the surface system in the subsurface matrix.

3.5.9. *Non-Orthogonality*

In Sec. 3.5.3, we described the use of stream lines for setting up the spatial discretisation. This system is useful only in so far as the flow is along the stream lines and such a grid system would not be chosen for a different application. This is because the choice of a non-orthogonal system should lead to transmissibility terms between cells not usually connected in the normal five point differencing schemes. However, for the stream line choice, the flux in these directions is zero so that neglecting these terms does not lead to errors.

For an application with reasonably fixed flow patterns (for instance the simulation of a symmetry element of a pattern development) this procedure can be very useful but care must be taken that field operating conditions do not seriously alter the stream lines.

Although distorted coordinate systems may give an improved geometrical representation of the reservoir and its fault system, it may also introduce a further level of divergence from the differential equation due to neglecting some cross-term transmissibilities. In this respect it is similar to an additional space truncation error.

The same is true of the case where local grid refinement is used as the fluxes between the fine grid area and the coarse grid area are subject to additional truncation error terms which depend on the degree of refinement.

However, apart from trying different grid systems (which is not frequently done), there is no *a priori* way of estimating the size of the error introduced.

A final comment on orthogonality of reservoir models is that the choice of axes, x along the dip, y along the strike and z vertical is non-orthogonal, except for horizontal models, and that the error introduced is usually small compared to all the other approximations used and the accuracy of the data available.

3.6. Rock Properties

The porous medium containing the hydrocarbons is a physically and chemically complex material whose properties and the dependence

of these properties on changes brought about by hydrocarbon production determine, to a large extent, the production characteristics.

In black oil reservoir simulation, most of these properties are
simplified or ignored, the rock being represented as a chemically inert,
temperature independent medium characterised by its porosity, permeability, rock compressibility and saturation dependent functions
of relative permeability and capillary pressure.

These properties are estimated from measurements made in a
variety of ways, including core data, interpretation of well test data,
interpretation of log data and the use of correlations.

3.6.1.   *Core Data*

The most direct measurements made are those made on the cores,
although even here, because the core has been brought to surface
and treated using some laboratory procedure, the results which are
obtained may be different from the *in situ* values.

Core analysis is broadly divided into the routine analysis,
performed rapidly soon after core retrieval, and the special core
analysis, which is much more time consuming, performed in order to
measure values of parameters used for petro-physical interpretation
and two-phase properties.

3.6.1.1.   *Routine core analysis*

From the point of view of constructing a reservoir simulation model,
one of the most important aspects is to decide upon the layering
system to be used. This usually depends on the vertical distribution
of both vertical and horizontal permeability and good core coverage
is a very desirable starting point as the routine analysis provides
numerical values for the porosity and permeability on some constant
sampling interval.

3.6.1.1.1.   Porosity

The porosity of interest to reservoir engineers is the value of
the connected porosity (which may be different from the absolute
porosity as measured by logs). This is rapidly measured by cleaning

and drying the core sample and comparing the weight of a helium container with and without the core. This gives a value for the grain volume and, knowing the value for the bulk volume, the connected porosity can be calculated.

An additional estimate of porosity is often available from the amount of fluids which were recovered from the core sample.

3.6.1.1.2. Permeability

The cleaned and dried core is inserted into a sleeve and a gas (usually air) is passed through the core. The pressure difference is adjusted to give a measurable flow rate from which the permeability can be calculated. The dependence of rate on permeability for a low pressure gas (density inversely proportional to pressure) becomes

$$q = \frac{kA}{2p_{\text{base}}L\mu}(p_1^2 - p_2^2), \tag{3.6.1}$$

where p_{base} is the base pressure and L is the length of the core plug.

The viscous flow of gases and liquids is not quite the same due to the different scale of the boundary layers. Klinkenberg showed that the gas permeability was related to the liquid permeability by

$$k_g = k_l\left(1.0 + \frac{c}{\bar{p}}\right), \tag{3.6.2}$$

where $c = $ constant and $\bar{p} = $ average pressure.

The liquid permeability can be obtained from the intercept of the plot of gas permeability versus the inverse of the average pressure. Usually the laboratory either quotes the necessary correction or actually applies it before reporting. The effect is usually to reduce the measured permeability by some 10%.

Darcy's law is valid for viscous flow, but if the flow rate is too high then an additional pressure drop due to turbulence results giving an apparent decrease in permeability with increasing rate. This means that permeability values in excess of 15,000 md are frequently not reliable but such values are usually isolated and infrequent.

3.6.1.2. *Geocellular up-scaling*

Simulators expect not the overall effective permeability but the average permeability of the net reservoir rock. Because of this, lateral transmissibility calculation includes the net-to-gross ratio and if this is already folded into the permeability up-scaling, the net-to-gross will be multiplied twice instead of only once.

Several methods can be used to up-scale the grid block parameters from fine scale geocellular models to coarser finite difference model grids. Great care needs to be exercised in scaling up the permeability in a way that is consistent with the way that the simulator uses net-to-gross ratio and permeability in order to calculate transmissibility values.

3.6.1.3. *Special core analysis*

The special core analysis experiments of interest here are those dealing with the pressure behaviour of the rock and those dealing with the effects of the presence of two immiscible phases.

3.6.1.3.1. Compressibility

In situ, rock is subjected to forces from the over burden and from the fluid occupying the pore space. The difference between these is the net confining pressure. As the pore pressure decreases, allowing rock grains to expand, the net confining pressure increases, tending to reduce the bulk volume of the rock. Both of these effects reduce the porosity and this effect is included in the formulation of the finite difference equations, usually as an elastic process with a constant compressibility.

Experimentally, the core sample is subjected to triaxial loading such that there is no deformation in the lateral direction and the change in volume is recorded as a function of the net confining pressure.

The values obtained should be used to compute the derivative of porosity with respect to pressure at the initial reservoir net confining pressure and corrected to give the uniaxial result by using the

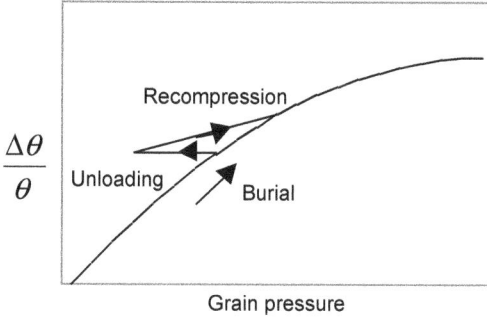

Figure 3.6.1. Rock compressibility.

correction factor:

$$f_{\text{corr}} = \frac{1}{3}\left(\frac{1+v}{1-v}\right)c_{\text{triaxial}}, \qquad (3.6.3)$$

where v is Poisson's ratio for the rock (approximately 0.3).

This gives a correction factor f_{corr} of approximately 0.6.

The effect of net confining pressure should be included in the values of porosity and (sometimes) permeability which are input to the model to correspond to the initial conditions.

If the reservoir to be simulated has a component of compaction drive then the effective rock compressibility can be an order of magnitude greater than the usual value ($3 \cdot 10^{-6}\,\text{psi}^{-1}$ to $7 \cdot 10^{-6}\,\text{psi}^{-1}$). The effect of compaction is not reversible so the standard elastic treatment is not sufficient and a special option dealing with this phenomenon should be used.

The effects of inelastic behaviour are demonstrated in Fig. 3.6.1.

3.6.1.3.2. Relative permeability

The presence of two immiscible fluids in the same pore space tends to alter the mobility such that a modification of Darcy's law is required. This is written

$$u(S_\Psi) = \frac{Akk_r(S_\Psi)}{\mu_\Psi}\frac{\partial}{\partial x}(-p + \rho_\Psi gz), \qquad (3.6.4)$$

where Ψ is the index of the phase and $k_r(S_\Psi)$ is the relative permeability which is a function of the phase saturation, varying between zero at the critical phase saturation to a value of one when the core is completely occupied by the phase.

On further examination, it is clear that this expression is a simplification as the behaviour of the fluid system when increasing the wetting phase saturation (imbibition) is not the same as when decreasing the saturation (drainage).

For practical purposes, the relative permeability of interest is that dealing with the displacement of oil by either water or gas. Typical oil–water and gas–oil relative permeability curves are shown in Figs. 3.6.2 and 3.6.3, respectively.

For water as the displacing phase in a water-wet rock this is the imbibition relative permeability having a dynamic range between the irreducible water saturation and the residual oil saturation.

This function is measured in the laboratory by either steady state or unsteady state measurements and can be performed at either laboratory or reservoir conditions.

The standard special core analysis for a water–oil relative permeability measurement is carried out using unsteady state displacement

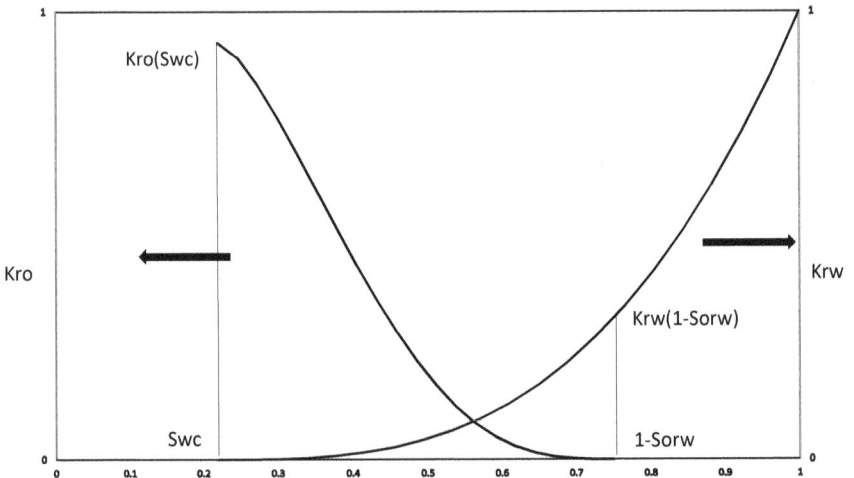

Figure 3.6.2. Water–oil relative permeability.

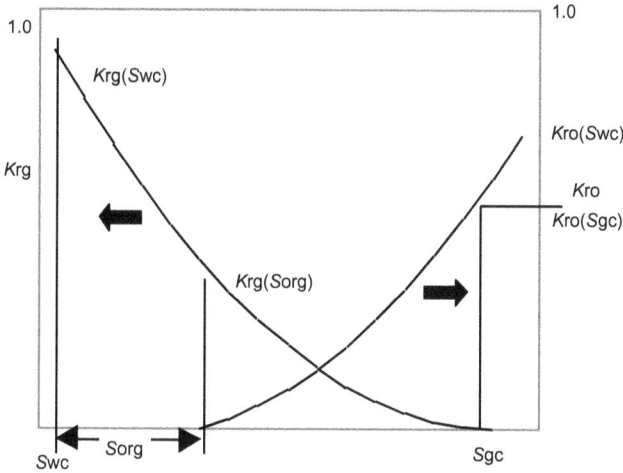

Figure 3.6.3. Gas–liquid relative permeability.

at laboratory conditions and the fluids used are water (or simulated formation water) and a standard mineral oil with a viscosity of approximately 20 centipoise.

This choice of viscosities permits the initial conditions of connate water saturation to be recreated in a short period of time and also produces rapid water breakthrough so that the full range of the relative permeability function can be measured.

This experiment is performed at high velocity in order to minimise the effects of the capillary pressure discontinuity at the entrance and exit faces of the core.

A more time consuming process is to measure the steady state relative permeability by simultaneous injection of water and oil into a core at rates corresponding to the velocity of fluids in the reservoir. In order to remove the capillary end effects, long cores can be used. This experiment may be carried out at laboratory conditions or at reservoir temperature and pressure with live crude and simulated formation water.

Because there is no longer a Buckley–Leverett displacement, the relative permeability function can be measured for the equilibrium core saturation which is set up as a result of the relative permeabilities.

This experiment, apart from difficulties with emulsion formation, requires time and a good deal of expense. However, there are indications that the residual oil obtained from such experiments is more representative for North Sea reservoirs than the values from the standard unsteady state method.

3.6.1.3.3. Capillary pressure

Capillary pressure arises because of the interfacial tension (σ) between two non miscible phases. If one phase wets idealised spherical rock grains in the presence of the non-wetting phase, then the pressure difference between the phases depends on the radii of the lenticular wetting phase volume. This is illustrated in Fig. 3.6.4 for idealised sand grains.

In the case of water and oil with water as the wetting phase, then

$$p_c = p_o - p_w, \tag{3.6.5}$$

where p_c is the capillary pressure given by

$$p_c = \sigma \left(\frac{1}{r_1} + \frac{1}{r_2} \right). \tag{3.6.6}$$

This gives rise to the characteristic capillary pressure curve as a function of wetting phase saturation. A typical capillary pressure curve is shown in Fig. 3.6.5.

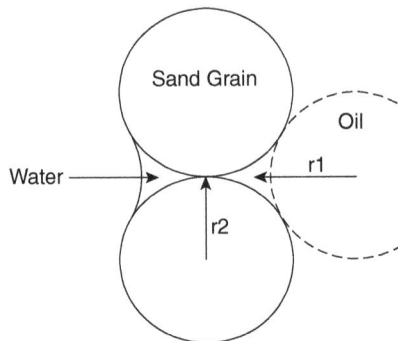

Figure 3.6.4. Origin of capillary pressure.

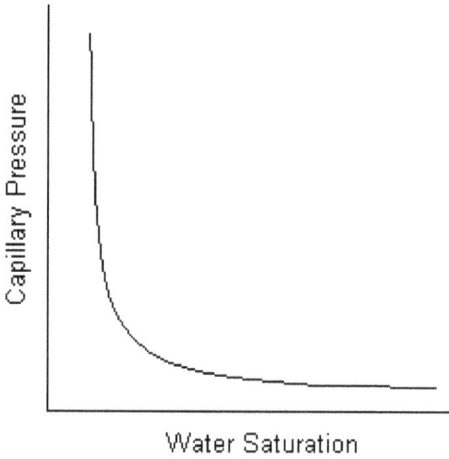

Figure 3.6.5. Capillary pressure.

The contact angle of the wetting phase actually depends on whether the wetting phase is advancing or retreating (see Fig. 3.6.6). This gives rise to a difference in the capillary pressure function for imbibition and drainage, with the drainage curve having a higher capillary pressure for any value of wetting phase saturation greater than the irreducible saturation.

This phenomenon, called hysteresis, affects both the capillary pressure, shown in Fig. 3.6.7, and the relative permeability, shown in Fig. 3.6.8.

As will be shown later, the capillary pressure is one of the determining factors in establishing an initial saturation distribution in the reservoir.

For practical purposes the data which are required correspond to the process of establishing the initial oil saturation which is usually considered to be that of oil migrating into the reservoir; a drainage process.

The capillary pressure then has a dynamic range between the irreducible water saturation and one.

In the laboratory, capillary pressure is measured by applying an external pressure and measuring the corresponding equilibrium saturations of the wetting and non-wetting phases. The fluid

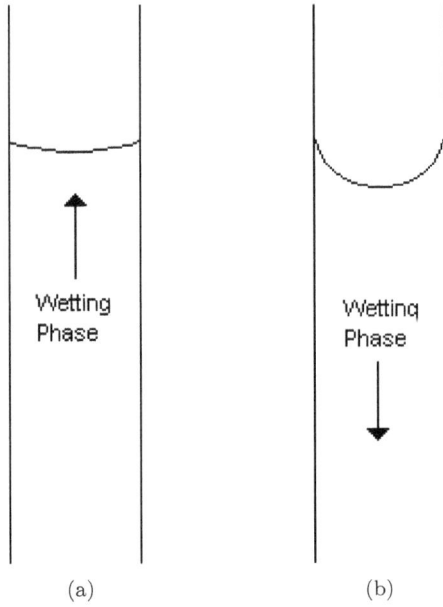

Figure 3.6.6. Contact angle for different flow direction.

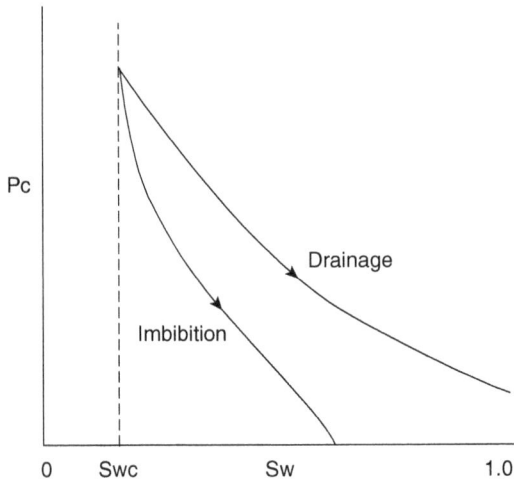

Figure 3.6.7. Hysteresis in capillary pressure.

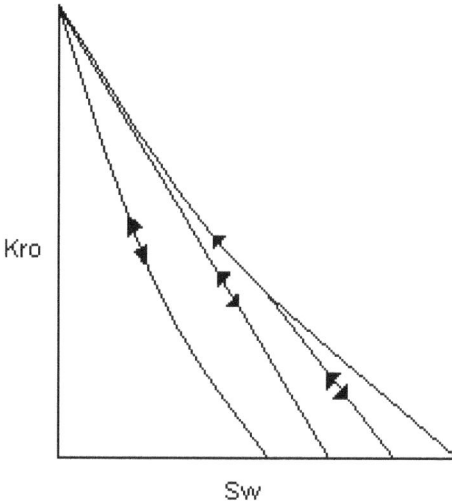

Figure 3.6.8. Hysteresis in relative permeability.

combinations normally used are an air brine system, where air is the non-wetting phase, or a mercury air system where air is the wetting phase.

In the air brine experiment, air is forced into the core reducing the initially high water saturation as the pressure increases. In the mercury air experiment, mercury is forced into the core plug and the saturation of mercury versus the mercury pressure is recorded.

The mercury injection experiment is frequently used as it is rapid to perform but it has the disadvantage that the core sample is permanently damaged and is of no further use.

3.6.2. *Log Data*

Electric wire line logs can measure quantities related to the *in situ* porosity and saturation. These can be used to give porosity and water saturation values and, in some cases, estimates of permeability.

An estimate of the porosity can be obtained from a comparison of the density and neutron tool response. The density tool relies on electromagnetic scattering of electrons in order to estimate the density as the electron density is closely related to the density of matter.

The neutron tool relies on the kinematic effect that the target with the greatest effect is one with the same mass as the projectile. This means that the neutron tool is sensitive to hydrogen atoms and hence to hydrocarbons and water.

A combination of these tools then, indicating a reduction in density and an increase in hydrogen content is indicative of a fluid filled porous rock and the excursion of the curves is related to the porosity and the fluid saturations. The response of the FDC–CNL log for a porous interval is shown in Fig. 3.6.9.

The fluid saturations may be estimated from the resistivity measurements because of the different electrical properties of drilling mud filtrate, formation brine and hydrocarbons.

Measurement of the resistivity deep in the formation away from the zone invaded by mud filtrate can give an estimate of the initial saturations and the shallow resistivity can give an estimate of the residual oil saturation. An example of the response to a water bearing formation is shown in Fig. 3.6.9.

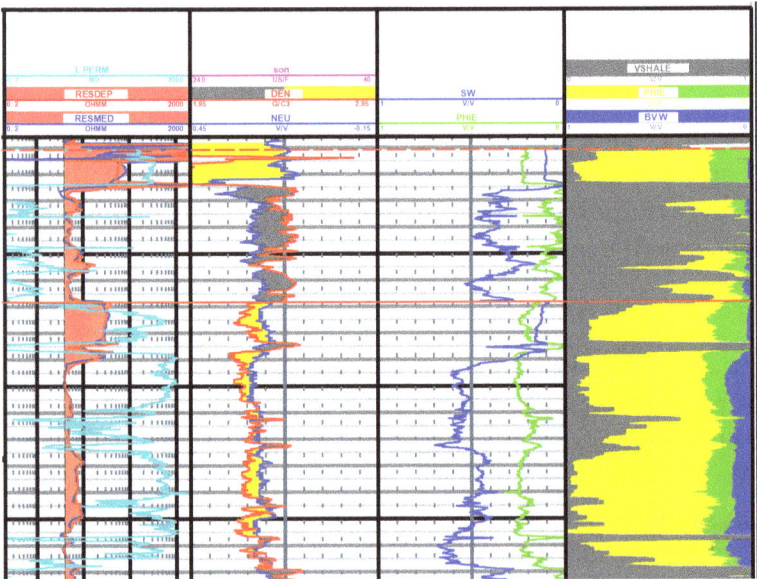

Figure 3.6.9. Neutron density and resistivity logs.

Wire line formation testing tools can, in principle, permit a measurement of the formation permeability by analysing the pressure build up in the sample chamber using theoretical solutions for spherical flow. Practically, these data, although immensely useful for the final pressure value, are little used for anything other than qualitative information regarding the formation permeability.

A more widely used method of deriving permeability from logs is to use a transform calibrated against core data from the field and geological interval, usually of the form:

$$k_{\log} = \alpha_{S_w V_{\text{clay}}} e^{\beta_{S_w V_{\text{clay}}} \phi}, \qquad (3.6.7)$$

where $\alpha_{S_w V_{\text{clay}}}$ and $\beta_{S_w V_{\text{clay}}}$ are the coefficients determined from the log and core data cross-plots obtained from wells where corresponding log and core data were available. An example of such a cross-plot is shown in Fig. 3.6.10.

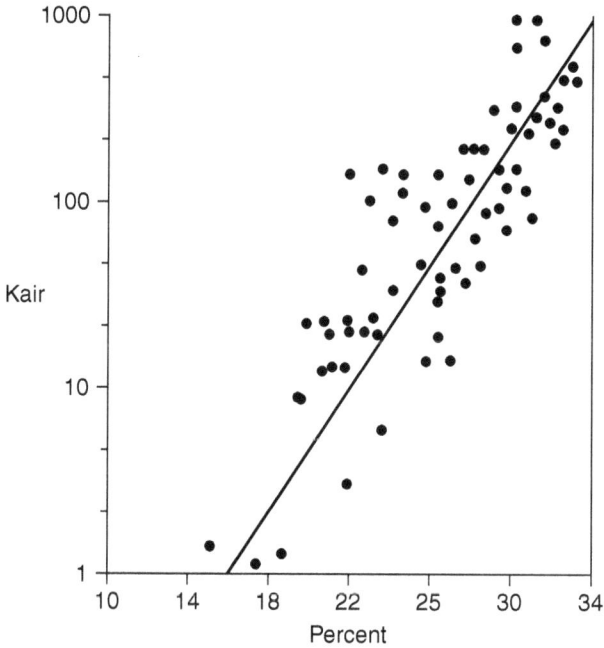

Figure 3.6.10. Porosity versus permeability.

3.6.3. *Test Data*

The permeability measurement from cores may suffer from effects due to the core handling and also represents a very small statistical sample compared to the volume of the reservoir.

By contrast, a well test is sensitive to the *in situ* permeability within the radius of investigation of the test and appears to overcome both of these difficulties.

For a well producing at constant rate in an infinite, homogeneous reservoir of constant thickness, a plot of the bottom hole flowing pressure versus the logarithm of time would give a straight line whose slope, m, is inversely proportional to the permeability

$$m = \frac{162.6 q \mu B_o}{kh}. \qquad (3.6.8)$$

In practice, the build-up is used more frequently than the draw-down data and both sets are frequently affected by afterflow, lateral changes in reservoir characteristics, communication to non-perforated zones and change of fluid type. A typical pressure build-up plot is shown in Fig. 3.6.11.

Because of this, the extraction of coherent data from a well test, as for the petro-physical data, is frequently a specialist task involving considerable interpretation. An analysis performed without consideration of possible variations in reservoir or fluid behaviour may be seriously in error.

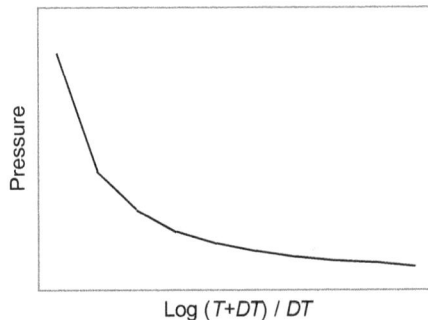

Figure 3.6.11. Typical pressure build up.

3.7. Model Relative Permeability

In Sec. 3.6, we discussed the measurement of relative permeability. The fact that such measurements are costly and time consuming means that the samples for which data exist might not have the required values of permeability and porosity for the model input. Also, the experiments excluded other effects, such as gravity, and this might or might not be representative of the actual displacement in the reservoir. This section describes the manipulation of the relative permeability and its use in developing pseudo relative permeability tables and the corresponding layering schemes.

3.7.1. *Data Manipulation*

The experimental procedure used to establish relative permeability relationships is specifically designed to exclude gravitational and capillary pressure effects. Their normal use is then for models where any such effects would be small within the grid cell which has the relative permeability value assigned to it.

 This implies that the cell size is small and in particular that the layer thickness is small which is the case for detailed single well models and for cross-sectional models.

 Here the requirement to minimise the number of grid cells to ensure that the problem is a tractable one is not usually a constraint so that the user is free to subdivide the vertical section into many layers so that cross-sections with hundreds of layers are not uncommon.

 Each layer has its permeability (horizontal and vertical), its net-to-gross ratio and its porosity assigned. The model also requires a relative permeability relationship (usually in the form of a table) but it is unlikely that for every layer a representative core sample will have been used for the experimental determination.

 The technique recommended is to carefully review the relative permeability data in order to select those experiments which appear to have consistent results and then to plot the normalised relative permeability curves.

This normalisation is carried out in two stages:

(1) the data are normalised over the dynamic range of the water saturation,
(2) the end point values of the relative permeability are normalised.

For the case of water–oil data, the relative permeability is first transferred to a new normalised saturation variable S_w^* where

$$S_w^* = \frac{S_w - S_{wc}}{1. - S_{wc} - S_{orw}},$$
$\qquad\qquad(3.7.1)$

The relative permeability values are then normalised according to the end point values as follows:

$$\bar{k}_{ro}(S_w^*) = \frac{k_{ro}(S_w^*)}{k_{ro}(S_{wc})}$$
$\qquad\qquad(3.7.2)$

and

$$\bar{k}_{rw}(S_w^*) = \frac{k_{rw}(S_w^*)}{k_{rw}(1. - S_{orw})}$$
$\qquad\qquad(3.7.3)$

The curves resulting from this procedure for the oil relative permeability are shown in Fig. 3.7.1.

Inspection of the resulting plots of normalised relative permeability permits one or more representative or average curves to be selected.

Different types of curve may be associated with say, different values of net-to-gross ratio.

In order to use the curves as input to the model, the end point values for the specific layer (or group of cells) must be specified. These values can be estimated from inspection of cross-plots of the relative permeability end points versus primary variables such as permeability and porosity.

Usually a simple set of correlations (or values) suffices to give a comprehensive set of relative permeability curves for the model e.g.

$$S_{wc} = \alpha - \beta\phi,$$
$\qquad\qquad(3.7.4)$

$$S_{orw} = \gamma,$$
$\qquad\qquad(3.7.5)$

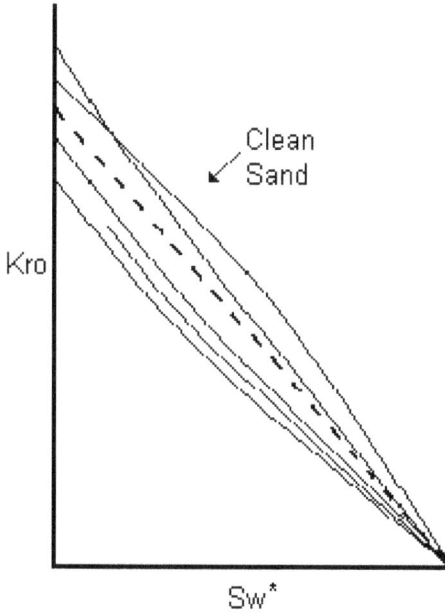

Figure 3.7.1. Normalised relative permeability.

$$k_{ro}(S_{wc}) = \delta + \log(k), \qquad (3.7.6)$$

$$k_{rw}(1. - S_{orw}) = \eta + \varepsilon\phi, \qquad (3.7.7)$$

Examples of cross-plots for connate water saturation and residual oil saturation are shown in Figs. 3.7.2 and 3.7.3, respectively.

These values can then be evaluated for each layer and the corresponding table of denormalised relative permeabilities constructed.

A similar exercise may be performed for the gas liquid relative permeability in order to obtain a corresponding set of tables. This table then represents the effective permeability of gas and oil as the oil saturation varies (keeping the water saturation fixed at the connate value).

A consistency requirement is that the corresponding oil relative permeability value (corresponding to $S_w = S_{wc}$ and $S_g = 0$) is the same for both tables.

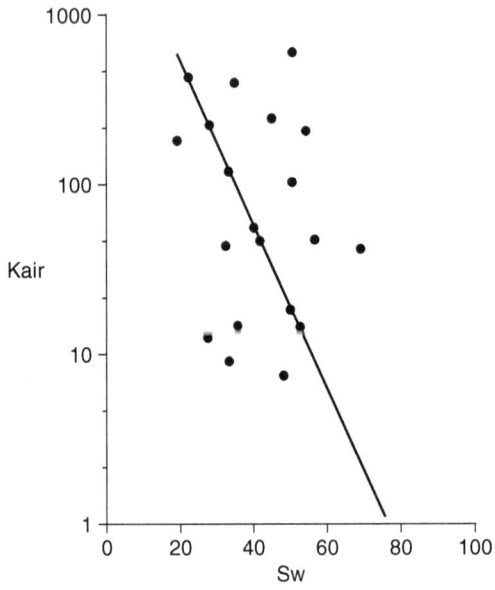

Figure 3.7.2. Water saturation versus permeability.

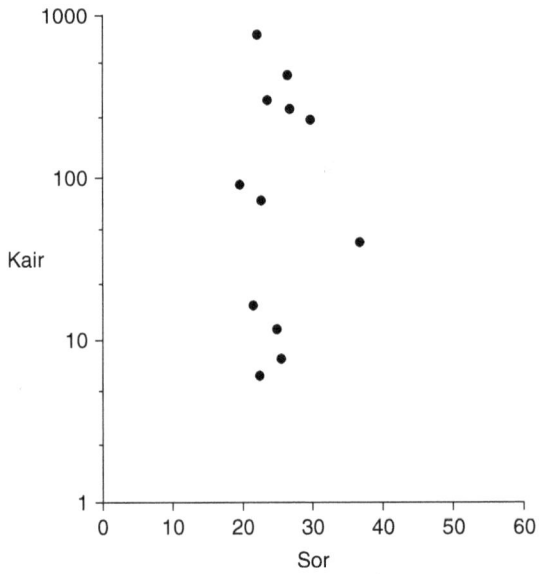

Figure 3.7.3. Residual saturation versus permeability.

3.7.2. *Three-Phase Relative Permeability*

Although the user enters only two tables each dealing with two-phase flow, in the model some cells might have all three phases mobile.

At this point models make use of correlations to compute the oil relative permeability, the most frequently used one being a modification of Stone's second correlation

$$k_{ro} = k_{ro}(S_{wc}) \left(\left(\frac{k_{row}}{k_{row}(S_{wc})} + k_{rw} \right) \right.$$
$$\left. \times \left(\frac{k_{rog}}{k_{ro}(S_{wc})} + k_{rg} \right) - k_{rw} - k_{rg} \right). \qquad (3.7.8)$$

An example of the behaviour of this correlation is shown in Fig. 3.7.4.

If all three phases have significant mobile saturations in a model, care should be taken that this correlation is producing the type of oil relative permeability expected. For most applications, gravitational segregation of the phases results in an extrapolation into the three-phase zone which is close to one axis so that the results are not usually sensitive to the form of the correlation used. This is not the

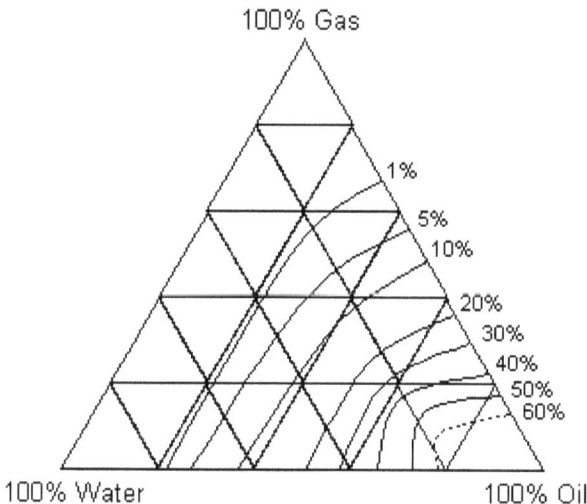

Figure 3.7.4. Three-phase relative permeability.

case for thermal simulation where there is close proximity of all three phases.

3.7.3. *Vertical Equilibrium*

If we consider a grid block with dimensions Δx in the x direction and Δz in the z direction, shown schematically in Fig. 3.7.5, the characteristic time for particle of fluid to traverse the block is:

$$\tau_r = \frac{\Delta x}{v_x} \qquad (3.7.9)$$

and

$$\tau_z = \frac{\Delta z}{v_z}. \qquad (3.7.10)$$

Flow across the block is determined by the viscous forces and flow down the block by gravity forces, so that

$$\tau_x \infty \frac{\Delta x}{k_x \frac{dp}{dx}} \qquad (3.7.11)$$

and

$$\tau_z \propto \frac{\Delta z}{k_z g \delta \rho}. \qquad (3.7.12)$$

If τ_x is much greater than τ_z then we can consider that within the block, gravitational equilibrium is established effectively, instantaneously giving fluid segregation within the block.

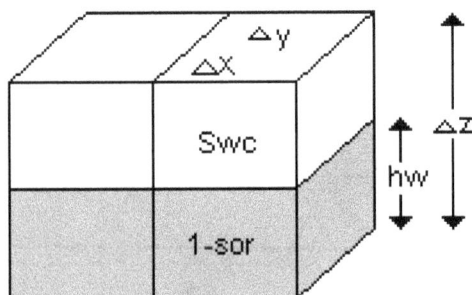

Figure 3.7.5. Schematic of vertical equilibrium cells.

For an oil-water system, using oil field units, this corresponds to

$$\frac{\Delta x}{\Delta z} \gg 10 \frac{dp/dx}{a}, \tag{3.7.13}$$

where a is the anisotropy ratio kv/kh.

If we assume this segregation occurs we can calculate the effective permeability for water and oil moving out through the side of the cell (not the top or bottom). Defining the transmissibility in the y direction as

$$T_y = \frac{\Delta x \Delta z}{\Delta y} k \tag{3.7.14}$$

then for water we have

$$
\begin{aligned}
T_{yw} &= T_y \bar{k}_{rw}(S_w) \\
&= \frac{\Delta x z_w}{\Delta y} k k_{rw}(1 - S_{orw})
\end{aligned} \tag{3.7.15}
$$

and for oil

$$
\begin{aligned}
T_{yo} &= T_y \bar{k}_{ro}(S_w) \\
&= \frac{\Delta x (\Delta z - z_w)}{\Delta y} k k_{ro}(S_{wc}).
\end{aligned} \tag{3.7.16}
$$

We can write the water saturation

$$S_w = S_{wc} + \frac{\Delta z_w}{\Delta z}(1. - S_{wc} - S_{orw}) \tag{3.7.17}$$

or

$$\frac{\Delta z_w}{\Delta z} = \frac{(S_w - S_{wc})}{(1. - S_{wc} - S_{orw})} \tag{3.7.18}$$

and substituting we obtain the effective water and oil relative permeabilities as functions of water saturation:

$$\bar{k}_{rw}(S_w) = \frac{S_w - S_{wc}}{1. - S_{wc} - S_{orw}} k_{rw}(1. - S_{orw}), \tag{3.7.19}$$

$$\bar{k}_{ro}(S_w) = \frac{1. - S_w - S_{orw}}{1. - S_{wc} - S_{orw}} k_{ro}(S_{wc}). \tag{3.7.20}$$

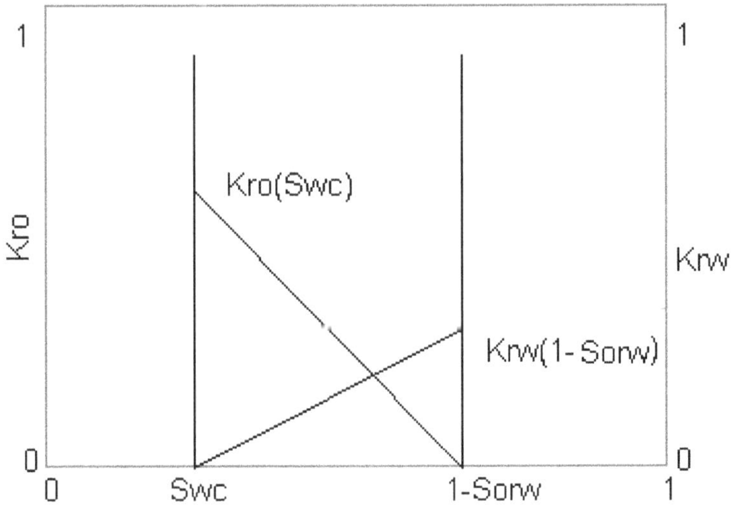

Figure 3.7.6. Vertical equilibrium relative permeability.

These expressions result in the linear relative permeability relationships, shown in Fig. 3.7.6, associated with the gravity segregation. The measured shape of the relative permeability, corresponding to the values at intermediate saturations is unimportant as, by assumption, these saturations never occur.

Thus we see, that given some assumptions regarding the spatial distribution of fluids in the grid block, we can compute a relative permeability which is different from that which is measured.

Conversely, if we input the modified relative permeability functions into the reservoir model, we are imposing the corresponding spatial distribution of fluids inside the grid block.

3.7.4. *Pseudo Relative Permeability*

In the previous section, we saw how simple assumptions regarding the dominance of one set of forces permitted an analytical calculation of the effective relative permeability functions.

Other examples of such calculations are the pseudo relative permeability functions which result from piston like displacement in a set of layers. If we have n such layers which can be referenced in the order in which they flood, then, when j layers have flooded, the relative permeability values and average saturations are given by

$$\overline{k}_{rw}(\overline{S}_w) = \frac{\sum\limits_{i=1}^{j} k^i h^i k^i_{rw}(1 - S^i_{orw})}{\sum\limits_{i=1}^{j} k^i h^i}, \qquad (3.7.21)$$

$$\overline{k}_{ro}(\overline{S}_w) = \frac{\sum\limits_{i=j+1}^{n} k^i h^i k^i_{ro}(S^i_{wc})}{\sum\limits_{i=1}^{n} k^i h^i}, \qquad (3.7.22)$$

where

$$\overline{S}_w = \frac{\sum\limits_{i=1}^{j} \phi^i h^i (1 - S^i_{orw}) + \sum\limits_{i=j+1}^{n} \phi^i h^i S^i_{wc}}{\sum\limits_{i=1}^{n} \phi^i h^i}. \qquad (3.7.23)$$

If we assume that the flooding order is according to decreasing values of permeability, then this gives Stile's formula for pseudo relative permeability. Conversely, if we assume flooding from the base upwards, this gives the Deitz pseudo curves.

In the case where no such simple assumptions can justifiably be made, then the results of a detailed cross-sectional model may be used to attempt to group together some layers and represent their joint behaviour by some pseudo function.

The existence of a coherent set of pseudo functions describing in a simple way the behaviour of a group of layers is one of the main bases for the vertical discretisation of the full field reservoir model, the other prime consideration being an adequate description of the perforated interval.

One simple method of calculating pseudo relative permeability from a cross-sectional model is, for each group of cells, to calculate the transmissibility weighted relative permeability. This is illustrated in Fig. 3.7.7.

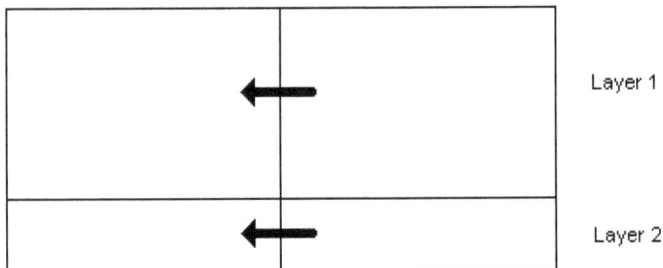

$$\overline{K}_{ro} = \frac{T_{r1}K_{ro1} + T_{x2}K_{ro2}}{T_{x1} + T_{x2}}$$

Figure 3.7.7. Pseudo relative permeability.

For a vertical column of cells

$$\overline{k}_{ro}(\overline{S}_{sw}) = \frac{\sum\limits_{i=1}^{n} T_x^i k_{ro}^i}{\sum\limits_{i=1}^{n} T_x^i}. \tag{3.7.24}$$

This formulation can be derived by making the assumption that the pressure gradient is the same in each layer, which is probably not correct. Similar assumptions exist in the formulation of other pseudo relative permeability functions. Because of this, all pseudo functions are approximations and the test for whether they are adequate is to resimulate the fine grid model with the coarse grid obtained from combining cells. If the results of the model are the same to within an acceptable (defined by the engineer) tolerance, then the pseudo functions may be used in subsequent coarse models.

The pseudo functions, which reflect the spatial distribution of the fluids, have been generated and tested with specific ratios between the viscous and gravity forces. If they are to be used in a full field model then the rate of production of reserves from the cross-section should be set to correspond to that expected from the field.

Conversely, if major changes in operating policy mean that this fractional off take changes significantly, it will be necessary to check that the pseudo functions are reasonable for the new velocities. If not, new functions should be generated and checked.

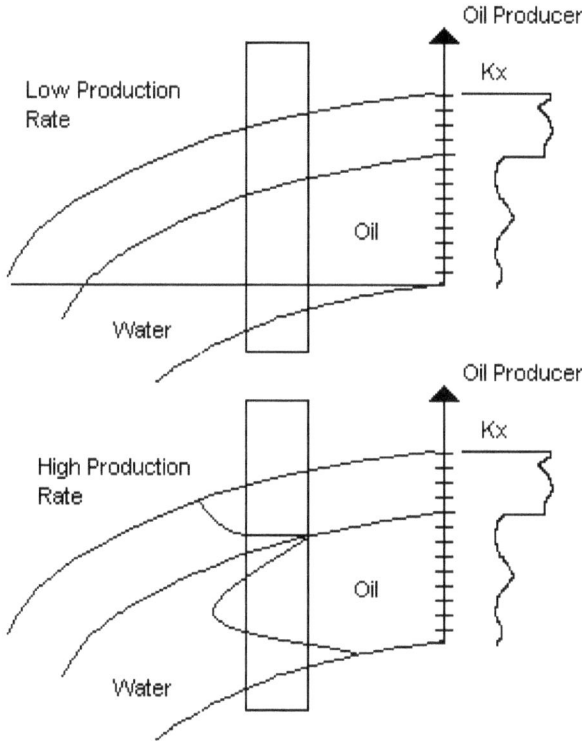

Figure 3.7.8. Rate dependence of pseudo curves.

The effect of modification of operating conditions on the location of fluids is shown schematically in Fig. 3.7.8.

The definition of the pseudo relative permeability used above was for combining cells in a vertical column but cells may also be combined in the longitudinal sense (see Fig. 3.7.9).

The calculation of the average saturation is straight forward but there is some choice for the selection of the relative permeability.

On the basis that transmissibility is truly related to the face between grid cells, the logical choice is the relative permeability of the cell closest to the outgoing face.

This gives rise to relative permeability functions shown in Fig. 3.7.10 which retard the advance of the displacing phase until the last block has a mobile displacing phase saturation.

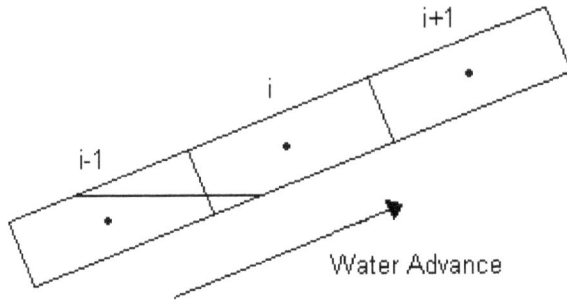

Figure 3.7.9. Longitudinal pseudos.

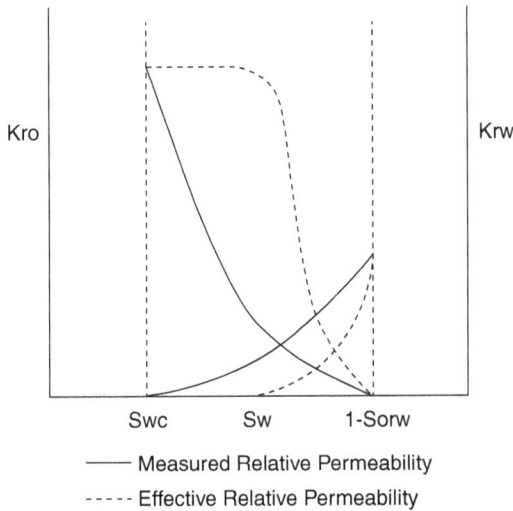

——— Measured Relative Permeability

- - - - - Effective Relative Permeability

Figure 3.7.10. Horizontal integration.

This process can be extended using the notion of Buckley–Leverett displacement so that the displacing phase is retarded until the shock front saturation is attained in the aggregate cell. The resulting relative permeability is shown in Fig. 3.7.11.

This leads to truncated relative permeability functions with very steep transitions corresponding to the change in mobility at the shock front. Such functions give severe problems for the Newton iteration schemes of fully implicit models and, if used, should be modified to have finite slopes.

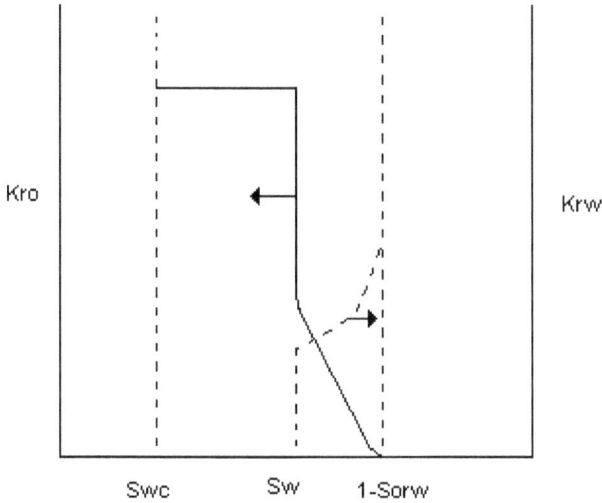

Figure 3.7.11. Truncated relative permeability.

These techniques can be used to limit the effects of numerical dispersion but can give numerical problems and also problems in dealing with flow reversal because the choice of the outgoing face is fixed.

3.7.5. *Well Pseudo Relative Permeability*

As well as describing the mobility of fluids in the reservoir as a function of the saturation values in the neighbourhood, the relative permeability functions relate the average saturation in a cell connected to a production well with the produced fluids. The behaviour of fluids flowing in the reservoir far from the well may not in any way describe the interaction of the phases in the vicinity of the well with the perforations.

A simple example of this is a well perforated over a small interval in a thick cell which has water and oil in vertical equilibrium. Until the water contact reaches the base of the perforations, no water is produced and when the contact reaches the top of the perforations, only water will be produced. Thus the linear behaviour of the relative permeability functions is reflected but over a reduced saturation interval compared to the function appropriate for cell-to-cell calculations.

Well pseudo curves may be used to approximate the effects of coning and cusping in coarse models which do not have sufficient spatial resolution to accurately simulate these effects.

3.7.6. *Summary*

The use of relative permeability arises because of the requirement to take into consideration the consequences of saturation dependent mobility at the microscopic scale.

These functions, when used on a macroscopic scale in reservoir models embody the microscopic effects compounded with assumptions regarding the distribution of reservoir properties and the spatial distribution of fluids within the finite difference cell.

The effects of the spatial distribution of fluids and the vertical distribution of horizontal permeability can be accounted for by using upscaling techniques which are representative of the large finite difference cells.

The grouping of cells is made by ensuring that the up-scaled functions used in the coarse model adequately represents the behaviour of the high resolution model.

3.8. Model Capillary Pressure

As described in Sec. 3.6, the capillary pressure is the difference in pressure between two immiscible fluids and is a function of the wetting phase saturation.

For a water-wet rock, the existence of a non-zero capillary pressure implies that the rock can maintain some water above the contact level where the water and oil phase pressures are equal. The magnitude of the water saturation at any point depends upon the balance between capillary and gravity forces.

The upwards capillary force is balanced by a downwards buoyancy force arising from the density difference of the two phases and the height above the free water level. For constant fluid density difference, the saturation, at any height above the contact, at which

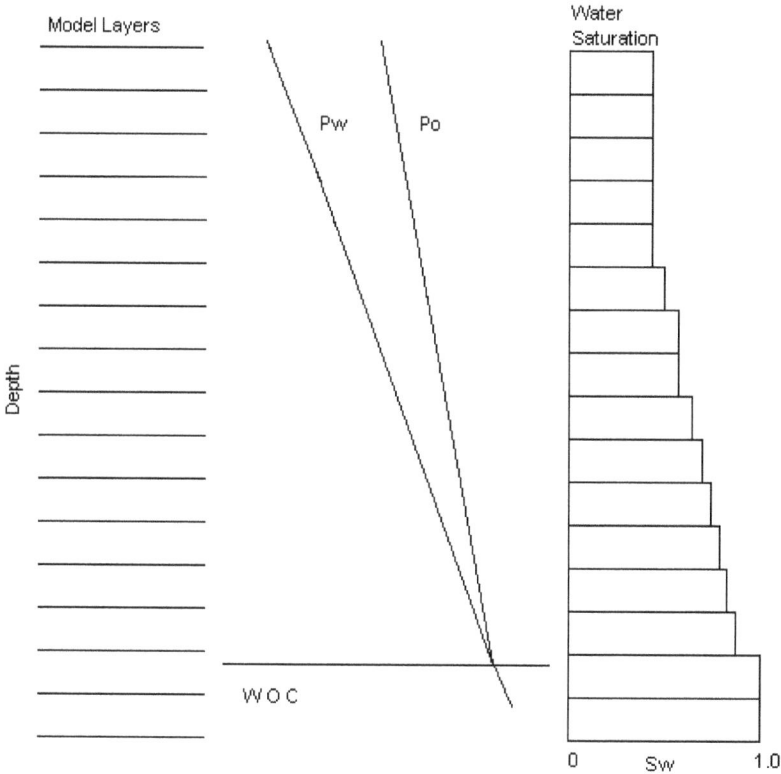

Figure 3.8.1. Model initialisation.

the capillary and gravity forces are balanced is a function of the
capillary pressure.

This gives rise to the transition zone of varying water saturation
observed in hydrocarbon reservoirs.

The technique used for initialising a reservoir model is shown
in Fig. 3.8.1. The procedure used is for each grid point (defined
by its cell centre) the downwards gravitational force, due to the
density difference times the height above the contact, is computed
and the corresponding equilibrium water saturation is obtained from
interpolation within the input capillary pressure data.

For cells which cut the contact there is the potential for errors
to occur as, if the cell centre is below the contact, the whole cell is
filled with water.

This can give rise to severe saturation errors at the edge of a model with thick grid blocks.

In order to overcome this problem, some models provide the facility of a more precise initialisation by subdividing the grid in the vertical sense by some factor. Each subdivided cell is then used as a calculation node for setting the saturation and the average saturation in the cell is computed in a way which is more consistent with the physical distribution of the fluids. This is known as the slice integration technique.

This can give rise to non-equilibrium terms at initial conditions as the inter block flow potentials are not computed using the same method but simply using the cell centres to compute the gravitational term and the average saturation for evaluating the capillary pressure curve.

For any non-zero and nonlinear capillary pressure function, the more precise initialisation will produce fluid flow potentials at initial conditions.

The size of these terms is usually small compared to the viscous pressure gradients set up by oil production but in special circumstances (thick grid blocks and high capillary pressure) it is advisable to check the movement by permitting some time steps prior to start of production.

3.8.1. *Manipulation of Capillary Pressure*

In the same way that it was necessary to find a representative relative permeability function for each rock type (see Sec. 3.7.1), the capillary pressure data can be normalised to find a representative or average curve.

The technique used is slightly different and is due to Leverett who recognised that capillary pressure is a function of the surface tension and the pore geometry. He used a relationship between the mean pore radius

$$\bar{r} \propto \sqrt{\left(\frac{\phi}{k}\right)} \qquad (3.8.1)$$

giving for the capillary pressure

$$p_c(S_w) = j(S_w)\sigma\sqrt{\left(\frac{\phi}{k}\right)}. \qquad (3.8.2)$$

The Leverett j curve then should be a function relating to a particular rock type and differences in porosity and permeability between the measured sample and the model layer are accounted for by the term $\sqrt{\phi/k}$.

This j function may be normalised over the dynamic range of the water saturation using

$$S_w^* = \frac{S_w - S_{wc}}{1. - S_{wc}}. \qquad (3.8.3)$$

An example of such a curve is shown in Fig. 3.8.2.

Note that the water saturation range is different to the normalisation for the relative permeability due to the difference in the direction of movement of the wetting phase. Here we are dealing with a drainage process whilst the relative permeability measurements correspond to imbibition.

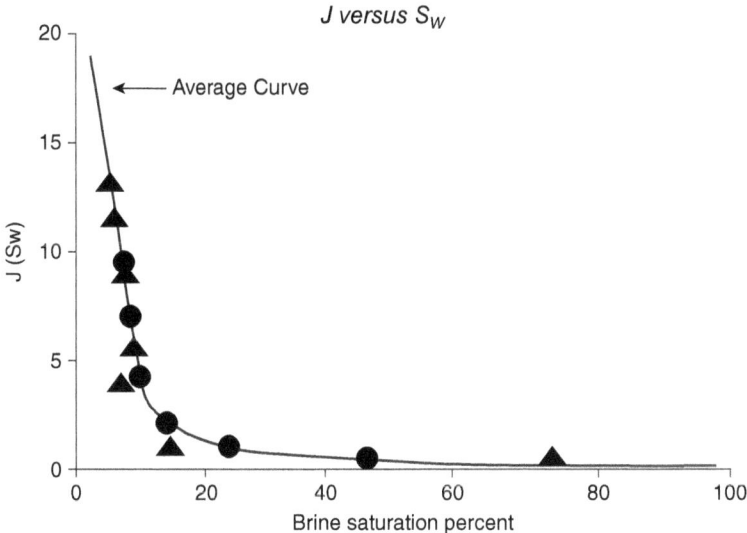

Figure 3.8.2. Typical leverett j curve.

For any model layer with porosity and permeability ϕ^i and k^i the j function can be used to compute the capillary pressure as:

$$p_c^i(S_w) = j^* \left(\frac{S_w - S_{wc}^i}{1. - S_{wc}^i} \right) \sigma_{res} \sqrt{\left(\frac{\phi^i}{k^i} \right)}. \qquad (3.8.4)$$

The capillary pressure function obtained in this way may be used in conjunction with the denormalised relative permeability functions giving a consistent saturation function for each model layer.

3.8.2. *Vertical Equilibrium*

In the same way that the vertical equilibrium relative permeability was developed in the previous section, we can derive the capillary pressure function for a block, initially filled with oil, through which the water contact rises.

Because the phase pressure difference is given at the finite difference node (the cell centre), if we consider a rising contact with no transition zone then for the contact at the base, middle and top of the cell we obtain the three capillary pressures:

$$p_{cb} = -\frac{1}{2}\Delta\rho g \Delta z, \qquad (3.8.5)$$

$$p_{cm} = 0, \qquad (3.8.6)$$

$$p_{ct} = +\frac{1}{2}\Delta\rho g \Delta z. \qquad (3.8.7)$$

The capillary pressure function which results is shown in Fig. 3.8.3 and, like the corresponding relative permeability, is a linear function of water saturation.

In order to obtain this relationship, we assumed that the transition zone was negligible compared to the grid block thickness. If this is not the case then we can compute pseudo capillary pressure functions by integrating the water saturation through the transition zone and relating this to the position of the water–oil contact.

As the contact moves then the phase pressure difference at the cell centre and the average saturation can be evaluated, giving a pseudo capillary pressure function. Because potentially this function

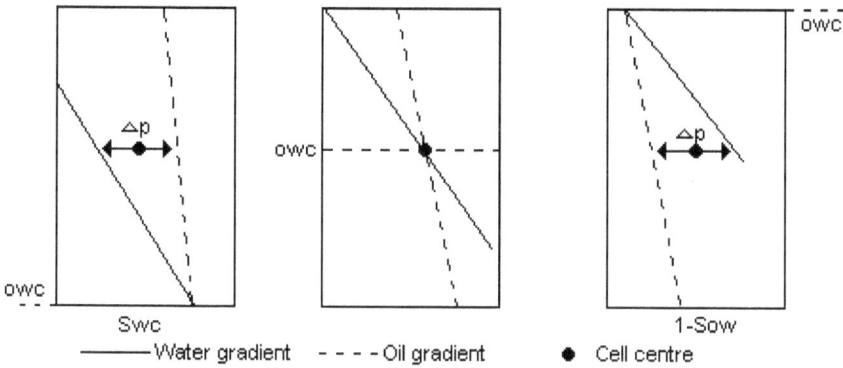

Figure 3.8.3. Vertical equilibrium pseudo capillary pressure.

is different for every cell, this approach, although giving the correct oil in place and stability at initialisation, is little used and has been superseded by the slice integration technique.

3.8.3. *Summary*

The concept of capillary pressure, as with relative permeability, arises from the microscopic effects of immiscible fluids occupying the same pore space. When used in a finite difference model, the capillary pressure. Along with the relative permeability, acquires a component which describes the spatial distribution of fluids inside the finite difference cell.

The user, by selecting different input for these functions can impose diffuse flow, segregated flow or whatever he believes most closely reproduces the reservoir behaviour, and to a large extent it is these assumptions and the resulting pseudo functions which determine the results of the model.

3.9. Fluid Properties and Experiments

Oil and gas reservoirs contain complex mixtures of hydrocarbon compounds together with non-hydrocarbon compounds at temperature and pressure conditions which are completely outside the everyday experience. The fluids are brought to the surface and processed through separators so that physically and chemically stabilised, non-toxic products can be sold. Some scientific approach to the understanding of the behaviour of the fluids can therefore have an impact on the quantity and quality of the sales stream, so that the study of hydrocarbon fluid behaviour has been pursued with much interest, leading to a large amount of data and theoretical models for describing their properties.

3.9.1. *Single Component Properties*

Prior to embarking on a description of the properties of hydrocarbon mixtures, it is useful to consider the simple case of a single component fluid system such as water.

It is well known that the boiling point of water is a function of the ambient pressure, and that this boiling point line in the pressure and temperature $(p-T)$ plane has a limited extension, beyond which no phase change occurs.

The line is called the boiling point line and the upper limit is called the critical point. At this point, the intensive properties of the liquid (those which are unrelated to the total mass) are identical to those of the vapour.

We can arrange to cross the boiling point line from liquid to vapour in many ways, but two simple ways are by fixing the pressure and increasing the temperature or alternatively, by fixing the temperature and decreasing the pressure.

The transition from one phase to another is not instantaneous, as the increase in temperature or decrease in pressure can only continue after all the liquid has vapourised.

At the boiling point line, although the temperature and pressure are held constant for some time, the fluid volume is experiencing extremely large changes.

3.9.2. *Properties of Mixtures*

If we extend the system so that we have two components, one of which is more volatile than the other, then the region in the $p-T$ plane where there is liquid vapour equilibrium is extended out from

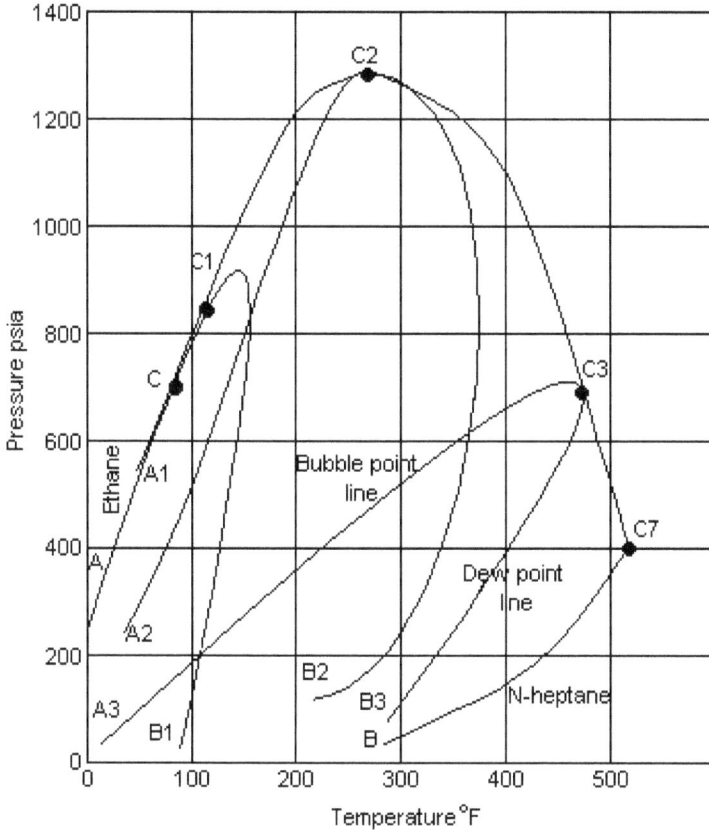

Figure 3.9.1. *P–T* Diagram for ethane *N*-heptane mixture.

the boiling point line to a two phase region. This is shown for a binary system of ethane and *N*-heptane in Fig. 3.9.1.

This two phase region also has limits in both temperature and pressure, but unlike the single component where the limits in temperature and pressure correspond to the critical point, the limits for a mixture are not coincident with the critical point.

As may be seen in Fig. 3.9.1, the behaviour of a mixture of two components is not a simple linear interpolation of properties and the critical pressure for the mixture may be significantly greater than that of either component.

As we cross the two-phase region from liquid to vapour, at constant temperature say, then the liquid saturation and the fraction of the volatile component in the liquid both decrease. The lines of equal liquid saturation (called isovols) or lines of equal composition (quality lines) can be drawn onto the two-phase region to give a quantitative expression of the change in the phase equilibrium across this region.

3.9.3. *Hydrocarbon Types*

The p–T diagram is a useful guide to the different types of hydro-carbons encountered. It should be recognised that the behaviour of a hydrocarbon mixture depends not only on the composition of the mixture but also on the temperature and pressure of its environment. This is illustrated in Fig. 3.9.2.

For example, a mixture which is a volatile oil at one temperature and pressure, would become a two-phase oil and gas system at lower temperature and pressure or a retrograde condensate at higher temperature and pressure.

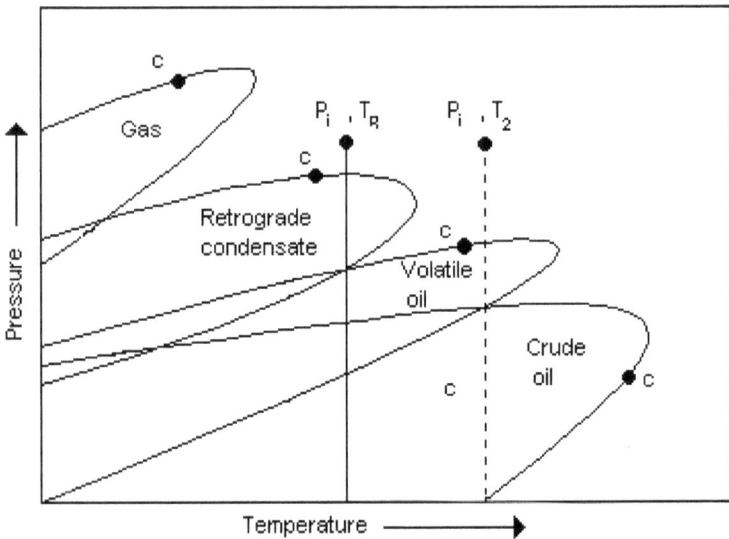

Figure 3.9.2. Effect of different fluids and temperature.

The classification of different hydrocarbon types is usually made on the gas–oil ratio, and this is related to the behaviour in the p–T plane as both temperature and pressure increase with increasing depth.

black oil	GOR $< 1,000$ scf/STB
high shrinkage oil	GOR $< 3,000$ scf/STB
volatile oil	GOR $< 5,000$ scf/STB
retrograde condensate	GOR $< 30,000$ scf/STB
wet gas	GOR $< 100,000$ scf/STB
dry gas	GOR $> 100,000$ scf/STB

The corresponding pressure versus temperature diagrams are shown in Fig. 3.9.3.

3.9.4. *Definitions*

Prior to discussing the experimental methods used to characterise hydrocarbon systems, we will define some of the more important terms used.

Saturation pressure

This is the pressure at which, during compression or expansion at constant temperature, the gas system is in equilibrium with an infinitesimal quantity of liquid or the liquid system is in equilibrium with an infinitesimal quantity of gas.

Bubble point pressure

For a specific temperature this is the saturation pressure where the liquid is in equilibrium with an infinitesimal quantity of gas.

Dew point pressure

For a specific temperature this is the pressure at which the gas is in equilibrium with an infinitesimal quantity of liquid.

Oil formation volume factor

The oil formation volume factor is the ratio of the volume in the reservoir at high temperature and pressure to the volume of the oil at

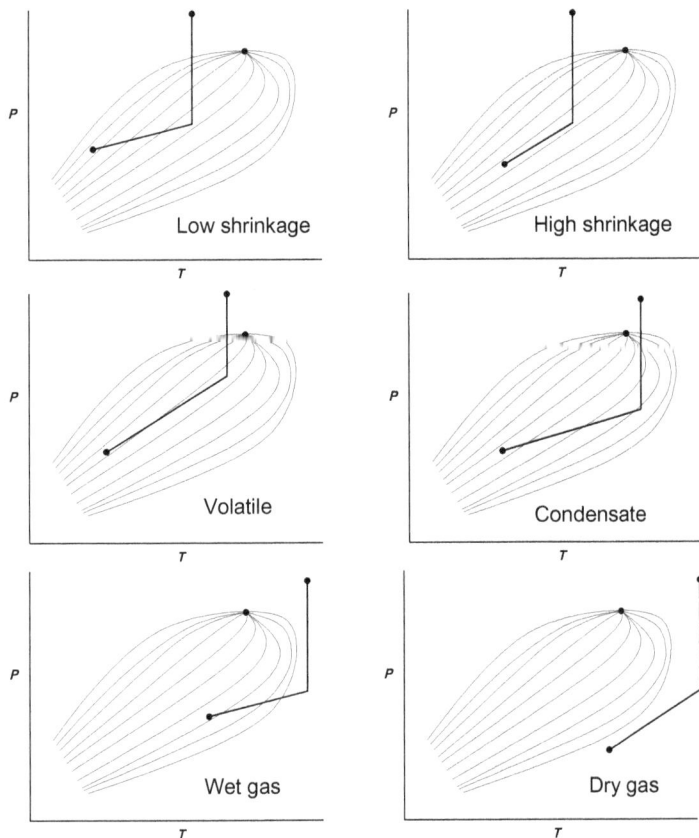

Figure 3.9.3. Fluid types.

standard conditions which results from producing the hydrocarbons to surface and passing them through the separation facilities.

Gas formation volume factor

This is the ratio of the volume of gas in the reservoir to the volume of the same mass of gas at standard conditions.

Gas deviation factor

The gas deviation factor is a factor used to modify the ideal gas law to take account of the finite size of and the interactions between gas molecules. For a given temperature, this relates the pressure volume product to the temperature value

$$pV = znRT. \tag{3.9.1}$$

The gas deviation factor and the gas formation volume factor are related by

$$B_g = \left(\frac{p_s T}{p T_s}\right) z, \tag{3.9.2}$$

where p_s and T_s are the pressure and temperature at standard conditions.

Solution gas content

This is the quantity of gas which can be dissolved in the oil at that specific temperature and pressure.

Oil gravity

This is the ratio of the oil density to the density of water. This is also quoted in terms of API gravity where:

$$\gamma_{\text{API}} = \left(\frac{141.5 \rho_{\text{water}}}{\rho_{\text{oil}}}\right) - 131.5. \tag{3.9.3}$$

Gas gravity

The gas gravity is the ratio of the density of the gas to the density of air at standard conditions. At low pressure, the deviation factor for both gases is approximately unity and so the gas gravity is given by

$$\gamma_g = \frac{M_g}{M_{\text{air}}} \tag{3.9.4}$$

where M is the molecular weight.

Coefficient of thermal expansion

This reflects the expansion of the fluid as a function of temperature.

$$c_{\text{Temp}} = \frac{1}{V}\frac{dV}{dT}. \tag{3.9.5}$$

Isothermal compressibility

The isothermal compressibility is the relative change in volume per unit pressure change:

$$c_{\text{oil}} = -\frac{1}{V}\frac{dV}{dp}. \qquad (3.9.6)$$

The following experiments set out to determine values for these quantities which specify to some reasonable accuracy, the behaviour of the hydrocarbon system.

3.9.5. *Experiments*

For a black oil system, the experiments are designed to provide a means of evaluating the pressure dependent fluid properties required in the formulation of models from material balance to fully implicit numerical models.

3.9.5.1. *Constant composition expansion*

The schematic experimental sequence is depicted in Fig. 3.9.4 where it can be seen that the sample is recombined at high pressure, and the temperature is raised to the reservoir temperature. During this phase, the coefficient of thermal expansion is measured. At a pressure above the reservoir pressure, the initial oil volume is measured and the pressure is reduced by increasing the cell volume by withdrawal of mercury. The oil volume is measured and the point at which bubbles evolve is noted.

The pressure is continuously decreased, keeping the evolved gas in contact with the oil. In some experiments, the oil volume is reported through the pressure drop cycle. The pressure is decreased until standard pressure is attained and the temperature is reduced to the standard value. The total amount of gas liberated and the gas gravity is recorded as well as the volume of the oil. The recordings taken during this experiment are indicated in Fig. 3.9.5.

3.9.5.2. *Differential liberation*

In the differential liberation experiment, the initial part is the same as for the constant composition expansion. The coefficient of thermal

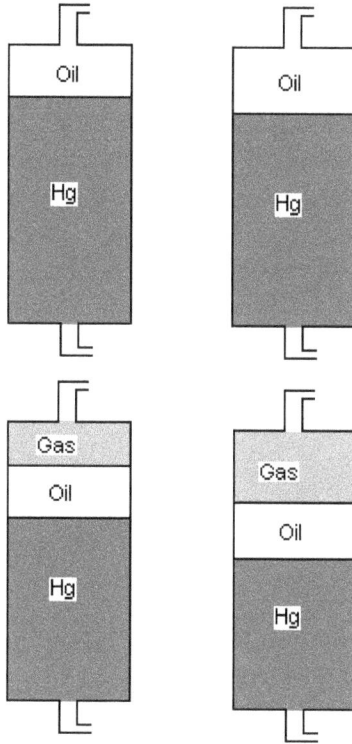

Figure 3.9.4. Constant composition expansion.

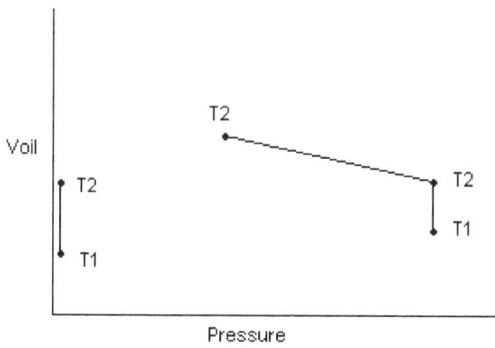

Figure 3.9.5. CCE results.

expansion and the fluid compressibility above saturation pressure are measured.

The procedure below the saturation pressure is to decrease the pressure in the cell from the saturation pressure to atmospheric pressure in 10 equal steps. At each step, the oil and gas system are allowed to come to equilibrium and all of the gas is displaced out of the cell. The volume of oil and gas are measured and the gas density is measured. This is illustrated in Fig. 3.9.6.

This process is repeated for each pressure in the series, so that at each pressure reading, the cumulative gas evolved and its density can be calculated. This experiment then gives volumes of the oil and

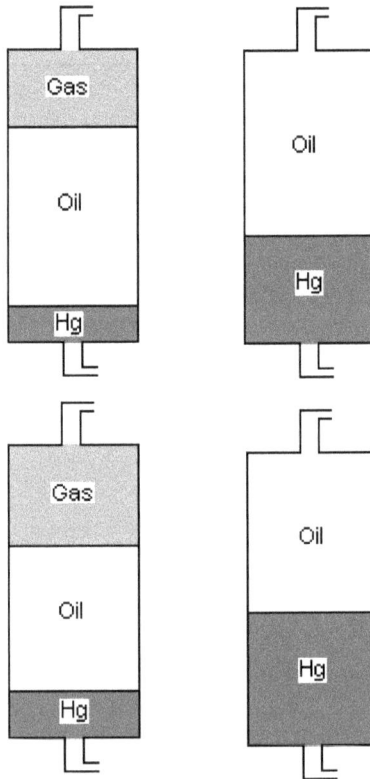

Figure 3.9.6. Differential liberation.

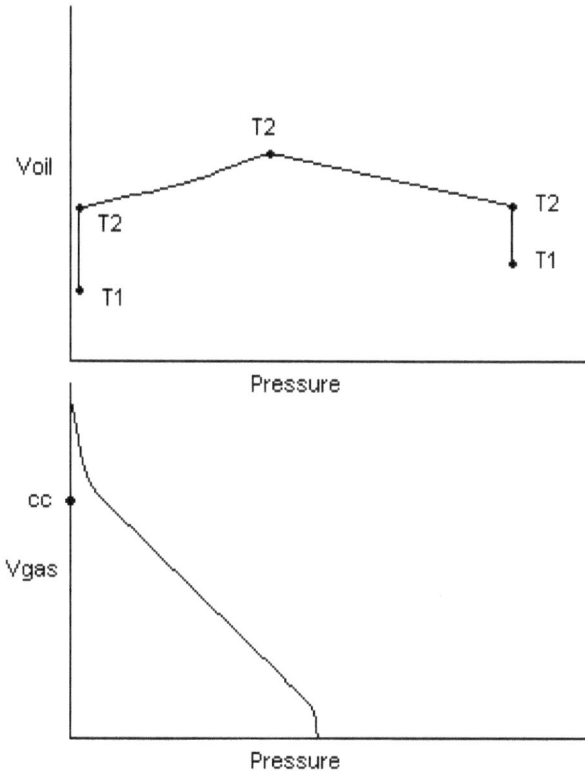

Figure 3.9.7. Differential liberation results.

gas below the bubble point as the pressure decreases. This is shown in Fig. 3.9.7.

In Sec. 3.9.2, we saw that the behaviour of a binary system in terms of the component properties was not at all linear. By performing the differential liberation experiment, at each pressure step we create a new mixture and it would be unlikely that the combined effect of the 10 pressure decrements would give the same result as the constant composition expansion.

In fact, the total amount of gas released (both mass and volume) tends to be larger for the differential liberation compared to the constant composition expansion.

3.9.5.3. *Constant volume depletion*

The constant volume depletion experiment consists of the initial part as for the constant composition expansion and differential liberation.

Below the bubble point, the pressure is decreased in a way similar to the differential liberation except that at every step, the gas is displaced until the initial cell volume is attained.

Thus the next pressure decrement is performed with the oil in contact with some of the gas from the previous steps, which alters the equilibrium compositions for the oil and gas.

The measured quantities for this experiment are the same as for the differential liberation but the quantity of gas released is usually smaller.

3.9.5.4. *Separator tests*

In order to better understand the relationship between the reservoir fluid and the sales stream, the reservoir fluid is subjected to simulated separation facilities.

This might consist of two pressure decrements and a final drop to standard conditions. During the pressure decrements, the temperature is also reduced so that the volume of gas evolved is minimised.

The tests will normally be performed for a range of separator conditions so that the optimum values in terms of retention of hydrocarbons in the liquid phase can be estimated.

3.10. Model Fluid Properties

From the laboratory data obtained, it is necessary to extract the information required to perform engineering calculations. The data requirements depend on the formulation of the calculations and below we describe the form of the data and the manipulation processes for the simplified case of a black oil model.

3.10.1. *Black Oil Fluid Properties*

Prior to discussing the black oil assumptions and their representation in a finite difference model, we will define some terms which are generally used in reservoir simulation models for black oil or more complex systems.

Component

This is a constituent of the hydrocarbon mixture which might correspond to a particular chemical species, a group of hydrocarbons within a range of carbon numbers or a statistical mixture of compounds evaluated by some means.

Phase

This refers to the state of a group of components which at the defined temperature and pressure form a coherent mixture with measurable properties. Thus in a saturated reservoir, the gas cap gas is an example of a phase.

Phase mole fraction of a component

This is the ratio of the number of moles of that component compared to the total number of moles of all components existing in that phase.

As a convention, the liquid mole fraction of a component i is written as xi and the gas mole fraction as yi.

Phase equilibrium

This is the state of dynamic equilibrium achieved when at a stable temperature and pressure, the compositions of the phases are independent of time, that is all values of x and y are constant.

At equilibrium, we can define a relationship between the x and y values for the various components as

$$y_i = K_i x_i, \tag{3.10.1}$$

where

$$K_i = K(T, p, x_i, x_j, \ldots) \tag{3.10.2}$$

and K is called the equilibrium coefficient.

The computation of liquid vapour equilibrium is mathematically difficult and in simulation of complex processes can consume a large fraction of the computer time.

Black oil

In a black oil model, it is assumed that the hydrocarbon system can be adequately represented by two components with a distribution of molecular weights. This distribution is that determined by the separation facilities so that the two components correspond to stock tank oil and gas (separator gas plus the gas evolved from the separator liquid in going to tank condition).

The equilibrium coefficient for these two components at a fixed reservoir temperature is assumed to be a function of pressure only.

In addition, all reservoir processes are assumed to be isothermal and the temperature effects in producing to surface are accounted for in the formation volume factor which relates surface to subsurface volumes.

An assumption which is usually made (but can be relaxed) is that the stock tank oil is non-volatile even at reservoir conditions and that it does not partition into the vapour phase.

This implies that the vapour mole fraction for the gas component is unity and that the reservoir gas has a fixed composition identical to the surface gas.

Together with the assumption regarding the equilibrium coefficient

$$y_{\text{gas}} = K(p)x_{\text{gas}} \qquad (3.10.3)$$

we see that the ratio of the number of moles of stock tank oil in the liquid phase to the number of moles of gas in the liquid phase is determined uniquely by the equilibrium coefficient, $K(p)$.

This means that for a given mixture of components at a set pressure, knowing the form of $K(p)$ (which is a transformation of the solution gas content), the liquid vapour equilibrium mixture can be computed *a priori* with no iterations.

Water is assumed not to partition into either hydrocarbon phases and not to influence any of the hydrocarbon properties.

The data required for a black oil model is then

(1) the oil formation volume factor versus pressure,
(2) the solution gas–oil ratio versus pressure,

(3) the oil viscosity versus pressure,
(4) the stock tank oil gravity,
(5) the gas formation volume factor versus pressure,
(6) the gas viscosity versus pressure,
(7) the gas gravity,
(8) the initial water formation volume factor and
(9) the water compressibility.

3.10.2. *Data Manipulation*

If the simulation is expected to enter the two (hydrocarbon) phase region, then the pressure dependence of the solution gas–oil ratio is required.

The experiment which gives estimates for this is the differential liberation experiment but this also tends to over estimate the total amount of gas liberated. (It accurately simulates a 10 stage separation at reservoir temperature).

The accepted approach is to use these data scaled to give the correct total gas–oil ratio and to assume that the form of the curve between the bubble point and the stock tank conditions scales in the same way.

The simple linear scale factor is defined as

$$R_s^*(p) = \left(\frac{R_s^f(p_b)}{B_{ob}^f} - \frac{R_s^d(p_b) - R_s^d(p)}{B_{ob}^d} \right) B_{ob}^f. \qquad (3.10.4)$$

Here superscripts refer to either flash or differential liberation and the subscript b indicates that these values correspond to the bubble point.

For consistency, this equation has been written using R_s as the volume of gas dissolved in the oil at that pressure. Laboratories frequently report the value which they measure which is the volume of gas (R^*) evolved from the oil. For the differential liberation experiment, this corresponds to the term

$$R_s^*(p) = R_s^d(p_b) - R_s^d(p). \qquad (3.10.5)$$

The formation volume factor is also scaled so that the bubble point value corresponds to the flash experiment.

$$B_o(p) = B_o^d(p)\frac{B_{ob}^f}{B_{ob}^d}.$$ (3.10.6)

Here, the correct values have been taken as those from the constant composition expansion (flash) experiment. If separator tests have been made it is preferable to use those which most closely resemble the field conditions.

After performing the scaling, we have corrected tables of data, which for oil properties typically appear as in Figs. 3.10.1–3.10.3.

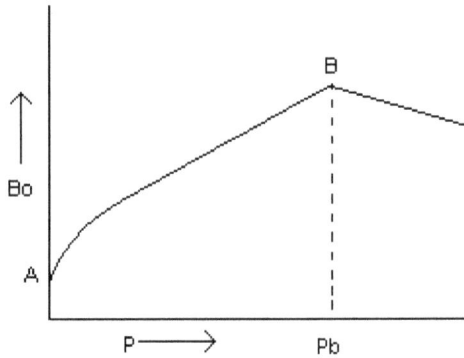

Figure 3.10.1. Oil formation volume factor.

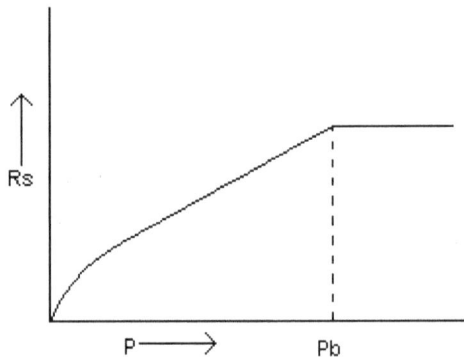

Figure 3.10.2. Solution gas–oil ratio.

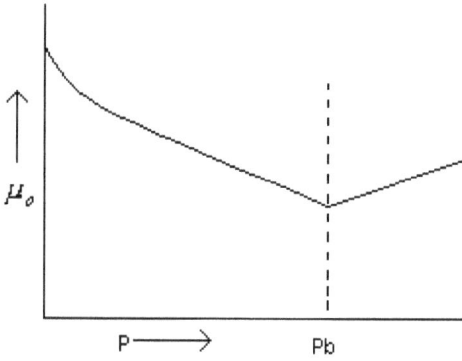

Figure 3.10.3. Oil viscosity.

These are however not in the form required by reservoir simulation models.

This is because of the requirement to treat the bubble point as a time dependent variable.

Consider an oil reservoir, which initially contains oil having the same properties everywhere, whose pressure falls below the bubble point and then rises again. Due to the different mobility and density of the liquid and vapour phases, some segregation might occur during the period of reduced pressure.

When the pressure rises, the pressure dependent equilibrium inherent in the model will cause the gas in the vapour phase to partition into the liquid phase raising the saturation pressure. This process ends either when all the gas is in the liquid phase or when the liquid phase is saturated.

For those cells at the base of the reservoir model which are deficient in gas the first case will occur and the model must compute under saturated values corresponding to a bubble point lower than the initial value.

If the pressure continues to rise, the gas rich cells will also reach saturation pressure and pass into the under-saturated region, requiring the model to compute first the saturated quantities and then the under-saturated quantities at bubble point pressures higher than the initial value.

The model then requires data which enable it to compute oil properties not only along the line defined by the modified differential liberation data, but for any value of bubble point pressure and reservoir pressure.

For this reason, the saturated solution gas–oil ratio and formation volume factor must be entered, extending above the expected maximum pressure.

The formation volume factor in the under saturated region is then calculated using a value of the oil compressibility or by interpolation between several sets of under-saturated data.

The treatment of the under-saturated viscosity is similar implying a requirement for a viscosity compressibility factor. The oil compressibility (if used) and viscosity compressibility are considered as constant.

Using a pressure p in the under saturated region, we can define the compressibilities

$$c_o = \frac{1}{B_{ob}} \frac{B_{ob} - B_o(p)}{p - p_b}, \qquad (3.10.7)$$

$$c_{\mu o} = \frac{1}{\mu_{ob}} \frac{\mu_o(p) - \mu_{ob}}{p - p_b}. \qquad (3.10.8)$$

For any new value of bubble point pressure, p_b^*, then we can write

$$B_o(p) = B_o(p^*)(1 - c_o(p - p^*)), \qquad (3.10.9)$$

$$\mu_o(p) = \mu_o(p^*)(1 - c_{\mu o}(p - p^*)). \qquad (3.10.10)$$

Such a treatment of fluid properties allows the bubble point to become a variable in the simulator.

3.10.3. *Spatial Variations*

The assumptions used to develop the black oil description appear to be extremely restrictive. Some flexibility is available in attempting to model spatial variations of fluid properties, but the two component formulation together with the requirement of equilibrium at initial conditions restrict such variations to be depth related.

The reason for this is that variation of fluid properties in a way not related to depth would lead to different fluid densities at the same depth, so that even if at that depth all fluid potentials were zero, at depths above and below, non-zero potentials would exist.

3.10.3.1. *Variable bubble point*

The initial composition of the oil is frequently a function of depth. Typically the bubble point pressure decreases with depth, which together with the increasing pressure gives a slight curvature to the pressure versus depth curve.

Because the pressure at a given depth depends on the integral of the fluid densities above it, the pressure is calculated on a regular and relatively fine vertical grid so that the system of equations converges rapidly.

3.10.3.2. *Variable Api gravity*

In a similar way, the density of the stock tank oil may increase with depth giving rise to a similar nonlinear pressure versus depth curve.

In both cases, the description of the spatial variation is made using an initially different mixture of components to describe the fluids. As long as the components are treated in the mass balance equations, then the initial variation will be reflected in the subsequent movement and production characteristics.

3.11. Aquifer Treatment

A primary source of reservoir energy for hydrocarbon production is the expansion of the water zone in contact with the reservoir. Because of the potentially large volume of water involved compared to the reservoir volume this expansion can be extremely important for the reservoir production characteristics.

In principle, the aquifer could be included in the reservoir model as part of the finite difference grid (see Fig. 3.11.1), but to do so would entail a large overhead in computing water flux within the (poorly defined) aquifer region when the main concern is only how much water flows into the reservoir during some time interval.

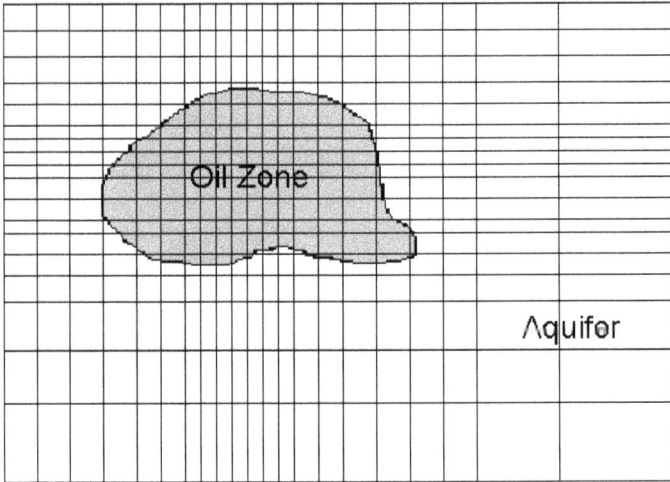

Figure 3.11.1. Aquifer in grid.

Various techniques have evolved to enable a realistic but efficient simulation of the aquifer response.

3.11.1. *Hurst Van Everdingen*

Hurst and van Everdingen used a modification of the solution of the radial form of Darcy's law used for well test analysis, solving for the rate by specifying a constant terminal pressure.

They then evaluated dimensionless water influx functions for infinite radial and linear aquifers.

For a constant terminal pressure, the cumulative water influx was given by

$$W_e = U \Delta p W_D(t_D), \tag{3.11.1}$$

where

W_e = cumulative water influx,
U = aquifer influx constant,
W_D = dimensionless water influx function,
t_D = dimensionless time.

The cumulative water influx, after $n+1$ time steps, for a cell coupled to the aquifer is

$$W_e(t_n + 1) = U \sum_{j=0}^{j=n} \Delta p_j W_D(t_{Dn+1} - t_{Dj}). \qquad (3.11.2)$$

This requires that for each cell coupled to the aquifer, the history of pressure changes must be stored and used to compute the current value of the influx.

Because of the computational overhead involved, more approximate methods have been developed.

3.11.2. *Carter Tracy*

Carter and Tracy approximated the water influx as a set of intervals with constant rate of water influx. The approximate difference in cumulative influx was written

$$W_e(t_{n+1}) - W_e(t_n) = (t_{Dn+1} - t_{Dn})$$
$$\left\{ \frac{u\Delta p_{n+1} - W_e(t_n)\left(\frac{dp_D}{dt_D}\right)_{n+1}}{p_D(t_{Dn+1}) - t_{Dn}\left(\frac{dp_D}{dt_D}\right)_{n+1}} \right\}, \qquad (3.11.3)$$

where

Δp_{n+1} = total pressure drop,
$p_D(t_D)$ = dimensionless pressure.

The dimensionless pressure function is the constant terminal rate solution to the radial form of the diffusivity equation.

3.11.3. *Fetkovitch*

Fetkovitch approximated the rate of influx into the reservoir using an analogy of the productivity index

$$\frac{dW_e}{dt} = J(\bar{p} - p_{n+1}), \qquad (3.11.4)$$

where

J = the Fetkovitch influx constant,

p = the average aquifer pressure,

p_{n+1} = the pressure of the cell.

For finite aquifers, the influx into the reservoir is taken into account to compute the average pressure by material balance for the next time step.

3.11.4. *Numerical Aquifer*

In order to try to model the transient pressure response of the aquifer without the need for a large number of cells, some models permit the user to define an auxiliary coarse 1D aquifer model which is coupled (using special transmissibility connections) to the main model. This is illustrated in Fig. 3.11.2.

This approach overcomes the problems of the approximations used for the analytical influx functions, but does leave the potential for including large space truncation errors into the model by coupling it to an extremely large grid block in the aquifer model.

To avoid this, the size of the grid blocks in the aquifer should be graded upwards in a smooth way from dimensions similar to those used in the reservoir zone.

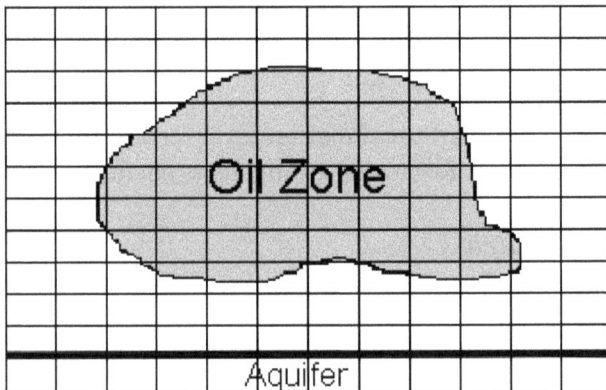

Figure 3.11.2. Numerical aquifer.

3.12.   Model Well and Production Data

Up to this point, with the exception of well pseudos, we have been concerned with modelling the properties of the fluid flow in the reservoir. In fact, this flow is initiated and sustained by production from wells and injection into wells and a realistic treatment of the behaviour in and around the well is a necessary step in the process of reservoir simulation.

3.12.1.   *Well Inflow*

For a well located in some drainage area with a radius of r_e and a pressure at the external radius of p_e, the operationally derived relationship for the flow rate of oil from the formation is

$$q_o = PI(p_e - p_{wf}). \tag{3.12.1}$$

From the solution of Darcy's equation, the productivity index, PI, can be expressed as

$$PI = \frac{2\pi k k_{ro} h}{\mu B_o \left(\ln \frac{r_e}{r_w} + S - 0.5 \right)} \tag{3.12.2}$$

for the case of semi steady state, and

$$PI = \frac{2\pi k k_{ro} h}{\mu B_o \left(\ln \frac{r_e}{r_w} + S \right)} \tag{3.12.3}$$

for the steady state solution.
These become

$$PI^* = \frac{2\pi k k_{ro} h}{\mu B_o \left(\ln \frac{r_e}{r_w} + S - 0.75 \right)} \tag{3.12.4}$$

for the case of semi-steady state and

$$PI^* = \frac{2\pi k k_{ro} h}{\mu B_o \left(\ln \frac{r_e}{r_w} + S - 0.5 \right)} \tag{3.12.5}$$

for steady state when the productivity index is expressed in terms of the average pressure, $\bar{p}$

$$q_o = PI^*(\bar{p} - p_{wf}) \tag{3.12.6}$$

From these equations we can see that the productivity index is dependent on the fluid saturation in the neighbourhood of the perforations, via the relative permeability term, and also dependent on the pressure via the viscosity and formation volume factor.

Further, the index which controls the flow of fluid from the formation into the wellbore depends on the distance from the well at which we reference the reservoir pressure.

For a well in a reservoir model, these effects are accounted for in the following manner.

The inflow to the well from a cell at a pressure p and water saturation S_w is written

$$q_o(p, S_w) = T_{\text{well}} M_o(p, S_w)(p - p_{wf} - \varepsilon), \tag{3.12.7}$$

where

$$
\begin{aligned}
q_o &= \text{the volumetric flow rate of oil,} \\
T_{\text{well}} &= \text{the well connection factor,} \\
M_o(p, S_w) &= \text{the oil mobility term,} \\
p_{wf} &= \text{the flowing bottom hole pressure,} \\
\varepsilon &= \text{the gravity correction term.}
\end{aligned}
$$

The well connection factor (or well index) is defined as

$$T_{\text{well}} = \frac{\alpha k h}{\ln \frac{r_o}{r_w} + S}, \tag{3.12.8}$$

where

$$
\begin{aligned}
\alpha &= \text{the unit conversion factor,} \\
kh &= \text{the permeability thickness,} \\
r_o &= \text{the pressure equivalent radius,} \\
r_w &= \text{the wellbore radius,} \\
S &= \text{the skin factor.}
\end{aligned}
$$

Here, the pressure equivalent radius is usually obtained from the publications by Peaceman, the current one being:

$$r_o = 0.28 \frac{\sqrt{\sqrt{\frac{k_y}{k_x}}\delta x^2 + \sqrt{\frac{k_x}{k_y}}\delta y^2}}{\sqrt{\sqrt{\frac{k_y}{k_x}}} + \sqrt{\sqrt{\frac{k_x}{k_y}}}}. \tag{3.12.9}$$

The mobility term, $M(p)$, contains the saturation and pressure dependent terms leaving the well connection factor as a constant for a given perforation. This mobility term is expressed as

$$M(p, S_w) = \frac{k_{ro}(S_w)}{\mu_o(p)B_o(p)}. \tag{3.12.10}$$

For a high productivity index well with a low draw down, the computed flow rate may be sensitive to the gravity correction, ε, which corrects from the well reference depth to the sand face depth.

Various methods are available for this calculation depending on the treatment of the multi-phase density and the dating of the calculations. For very high deliverability wells, the results should be checked for sensitivity to this correction.

3.12.2. *Production Control Data*

The reservoir engineer performing a simulation study is faced with the task of representing the observed production data to the simulator, in order to try to match the reservoir behaviour, and then to attempt to predict and optimise the likely future performance of the reservoir taking account of the facilities and operational constraints. As each field is unique, the task of providing the engineer with a sensible but manageable choice of options has been one of the more difficult problems of practical reservoir simulation.

3.12.2.1. *Targets*

Target rates are those rates which the simulator should attempt to maintain provided that this can be achieved within the set operating constraints.

Rates may be specified in a variety of ways for different purposes. Examples are surface oil rate, gas rate, water rate, liquid rate or subsurface oil rate or total fluid rate. For injection wells, additional target rates can be given expressing the required rate as a function of the phase production (for example reinjecting 85% of produced gas) or reservoir voidage (for example total injection rate balancing the voidage rate) to give, to first order, a stable average reservoir pressure.

Inspection of the productivity index equations above, reveals that they are written for the oil phase. Similar expressions exist for the water and gas phases (although the gas equation may be more complex due to rate dependent skin factor and the difference in pressure behaviour of gases leading to the equation being linear in the real gas pseudo pressure).

This implies that if the user requests a well to produce a certain rate of oil per day, then whatever the ratio of the mobility terms for the water and oil phases, the well will produce fluids to give that water–oil ratio unless some constraint is activated.

This might be necessary as the well might not be capable of flowing at the simulated bottom hole pressure and water cut.

3.12.2.2. *Constraints*

Because a well in the simulation model could produce fluids in a way impossible for a real well, the user can elect to choose some constraints which can be applied.

These constraints may be related to wells, (for example a minimum tubing head pressure for all wells corresponding to the pressure needed to deliver hydrocarbons to the separator) or to groups of wells (a group of wells producing into one separator has a maximum group liquid rate determined by the separator size and retention time).

The constraints can be connected in a tree structure so that at each node (well, platform and field) constraints are honoured.

As well as gas–oil ratio limits and water cut limits, secondary rate limits can be set which will permit the primary target rate to be met provided that the secondary limit is not violated. This can be useful where there are maximum through put restrictions such as produced water disposal or gas treatment.

A useful constraint to activate is the economic limit which shuts in wells or terminates the simulation when lower limits for production rates are crossed.

In terms of oil field operation, if a well has some production problem then frequently the production is made up either by working over the existing well or by drilling a new well.

Reservoir models have a user definable set of actions which can be performed automatically in the model to attempt to simulate reasonable oil field practice in maintaining production levels.

3.12.2.3. *Actions*

Typically these actions consist of attempting to control the production of unwanted components such as water and gas by either altering the well rate (reducing the rate of high water cut wells), performing a work over on the well to squeeze off the worst offending perforation (this has a 100% success rate in models) or the well can be shut in.

If a group of wells has spare production capacity, then at some predefined drilling rate, wells can be added in specified locations in order to increase the production rate to correspond to the available facilities.

3.12.3. *Practical Considerations*

Full field reservoir simulation, which is relevant for the constraints and actions discussed above, is often performed in the stages corresponding to a refinement of reservoir description by matching the observed behaviour followed by a series of prediction runs which attempt to optimise the reservoir performance in some way.

These two modes of operation are very different in the way in which target rates and constraints are specified.

3.12.3.1. *History match*

During the period of a history match, oil gas and water rates will usually be available for each well on some reasonably frequent time basis.

In addition, down-hole reservoir pressure data at specific well locations will also be available.

The engineer performing the simulation study has to attempt to match the pressure behaviour and water movement in the reservoir from the well control data at his disposal.

If the input data to the model is specified in terms of produced oil rate, then for wells where the water cut is poorly simulated, the total volumetric production from the reservoir may be seriously in error giving rise to spurious pressure behaviour.

Also, the inclusion of (sensible) well constraints can have adverse consequences as for a well where the water cut is too high in the model, the well will be shut in where as it continued in production in the field.

Thus the rate specification and the inclusion of well constraints can lead to the incorrect volumetric fluid withdrawal making the task of achieving a history match almost impossible.

The problem becomes significantly more tractable if the well constraints are deactivated and the production rate data is converted to the equivalent volume of reservoir fluid at the ambient pressure, and this rate is produced regardless of the proportion of each component.

In approaching the simulation in this way, the problems of matching the pressure and water cut behaviour or gas–oil ratio behaviour are largely decoupled.

3.12.3.2. *Prediction*

During the prediction phase it is necessary that the production constraints such as minimum tubing head pressure and maximum liquid through put are implemented.

Figure 3.12.1 shows typical tubing lift curves for different water oil ratios. If several layers are perforated each having different water saturation, changing the bottom hole pressure can modify the water cut and the convergent bottom hole (and therefore tubing head) pressure must be found by an iterative procedure.

As in most iterative schemes, the closer the start value is to the solution, the faster the convergence and so during a prediction it

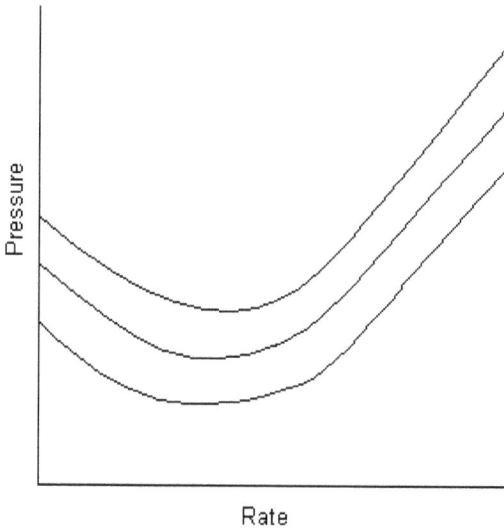

Figure 3.12.1. Tubing lift curves.

is worth while setting target rates which are close to the expected production rate.

It is a measure of the quality of the history match that the model should proceed from the history match to the prediction phase without experiencing instability or discontinuity.

Chapter 4

History Matching

Deryck Bond

Senior Consultant for Kuwait Oil Company

4.1. Introduction

It may be appropriate to start by attempting to define "history matching" in the context of reservoir modelling. One possible definition, variants of which are commonly used, is as follows:

> "History matching is the process of modifying the model input data until a reasonable comparison is made with historical data."

In this definition the emphasis is on producing a model where there is agreement between simulated and historical data. An alternative definition that may be preferable is

> "History matching is the process of making reasonable changes to model input parameters to better match historical data and to improve the predictive capability of the model."

This latter gives emphasis to the business aim of our work: producing a model that gives reasonable predictions of future performance that can influence our business decisions.

Either definition raises a number of questions, such as:

- How well do we need to match historical data?
- What changes to our original model are reasonable?
- How do we go about making suitable changes?
- Are there multiple ways we can match the observed data?

- How are our conclusions influenced by the limitations of the simulation tools we have?
- How can we form a view on the predictive capabilities of our model?
- Will a single (history-matched) mode allow us to meet our objective or do we need to develop multiple models?
- What if we cannot achieve a reasonable match? Has an exercise that has just invalidated a model been of any value?

These questions are common to modelling many "inverse problems" — problems where we wish to match model "outputs" (which may have significant uncertainty) by changing inputs (about which we have limited information). Other inverse problems in the petroleum industry are encountered in pressure transient analysis and in seismic interpretation.

History matching reservoir models is a relatively complex inverse problem. The underlying equations are non-linear; the tools used to solve these equations may have limited accuracy and the physical system is complex. Finding solutions to these types of problems is an active area of mathematical research. For any inverse problems there are a number of concerns related to how "well-posed" the problem is. These include the following:

(1) Is there a solution?
    Given that we are attempting to model historical data it would seem reasonable that there would be a solution. However there may be concerns about the limitations of the simulation model; if the simulation model has insufficient numerical resolution then we could envisage it not being possible to match the observed date with the simulation model.
(2) Is the solution unique?
    We shall see through examples that this is a problem that has practical implications for us. If we can match observations in many ways, then how much confidence do we have in our ability to predict?
(3) How "smoothly" should we model outputs depends on model inputs?
    This has clear implications for us being able to match the model.

Before attempting to address the above questions, we will put the process of history matching into an appropriate context. Firstly, it is necessary to put the exercise (and the wider reservoir modelling exercise that it is part of) into an appropriate business context; secondly, it is important to have a view of the history matching process as a part of the wider reservoir modelling work flow.

This is followed by a description of the work that would be involved in a "conventional" history matching exercise involving matching a single model of a reservoir and its use for prediction. This type of approach is summarised in the literature of Mattax and Dalton (1990). In doing this, emphasis is placed on some of the "practical" elements of the study, particularly on issues related to surveillance data, data acquisition, and quality control (QC) and "calibration" of the matched model to field performance. A process of matching the data is described that could use "manual" or "computer-assisted" approaches to altering model parameters. Problems associated with non-uniqueness are discussed.

Finally, comments and some references to automatic history matching and to the use of multiple models for uncertainty assessment are provided. These notes do not do justice to these very important issues.

4.2. Context of the History Match Study

As for any study, it is important to understand the role that the history match study plays in the wider context of achieving business goals, its relation to choices about reservoir management, and its relationship to the wider range of technical work that would be involved in a geological and simulation study.

4.2.1. *The Business Context of the Study*

For anyone involved in a history match exercise, having (and keeping throughout the study) a clear understanding of the business context of the study is vital. This is because the criteria for judging the "fitness for purpose" of the matched model will be a function of the business aims. Some examples can illustrate this.

Example 1 — A study of potential in-fill locations in a large mature field with a large volume of data. In this case, success would be judged by the ability to guide the choice of well location. The match to production and surveillance data, especially close to any proposed in-fill locations, would be expected to be good.

Example 2 — A study early in the life of a field intended to identify the time when artificial lift may be required. In this case success would depend on the ability of the model to predict factors that would influence the need for artificial lift (pressure decline and the tendency of well water cuts to rise). Achieving a good match to individual well data may be less important than in Example 1. On the other hand, assessing uncertainty may be more important; the business decision could be based on an estimate of the earliest that artificial lift would be needed.

Example 3 — A study aimed at predicting short- to medium-term production. This may be important to financial planning. In this type of application there would be a particular interest in ensuring that well performance was well matched at the start of the prediction period.

The scope for having multiple models in the above cases differs. In the first example the appropriate style of model (lots of history and lots of data) may be relatively deterministic. This, and the increased efforts needed to history match a field with a lot of data, could make having a single model a sensible choice. For the second example, the emphasis on assessing uncertainty and the reduced effort that may be associated with matching models with less data may make using multiple models both more useful and more practical.

4.2.2. *Relation to Reservoir Development/Management*

The reservoir studies of which history matching forms a part cannot be viewed in isolation from the wider subject of reservoir management. The ability to develop history-matched models that are useful in reservoir management may be limited by decisions that are made relating to how to develop the reservoir, what data to acquire, and how to analyse and manage those data.

The following can reduce our ability to produce a useful history match:

- Choosing to have commingled production from different "flow zones" (or even different reservoirs) without making efforts (such as production logging or using down hole measurements) to measure such production.
- Having wells that are in poor mechanical condition and can allow cross-flow between different sets of perforated intervals and/or flow behind pipe.
- Having limited reservoir characterisation and surveillance data (for example limited core data, the choice not to have 4D seismic data, infrequent bottom hole pressure measurements, limited fluid sampling, etc.).
- Having poor production (or injection) allocation because wells are not tested frequently enough.
- Not running Repeat Formation Tester (RFT) (RFT — a Schlumberger trademark — or similar tool) or similar logs on development wells.
- Failing to invest in good data management.

In all the above, there is a choice between generally lower- and higher-cost approaches to managing a reservoir. (Coring wells costs money; commingled production can save a lot of money.)

It is important in the context of a history match that the role of such decisions is recognised. Expectations about the benefits of a history match exercise will depend on previous reservoir management decisions.

4.2.3. *The Work Flow Context of the Study*

At this point, it is as useful to provide a very high-level view for history matching. The following flow diagram attempts to illustrate this.

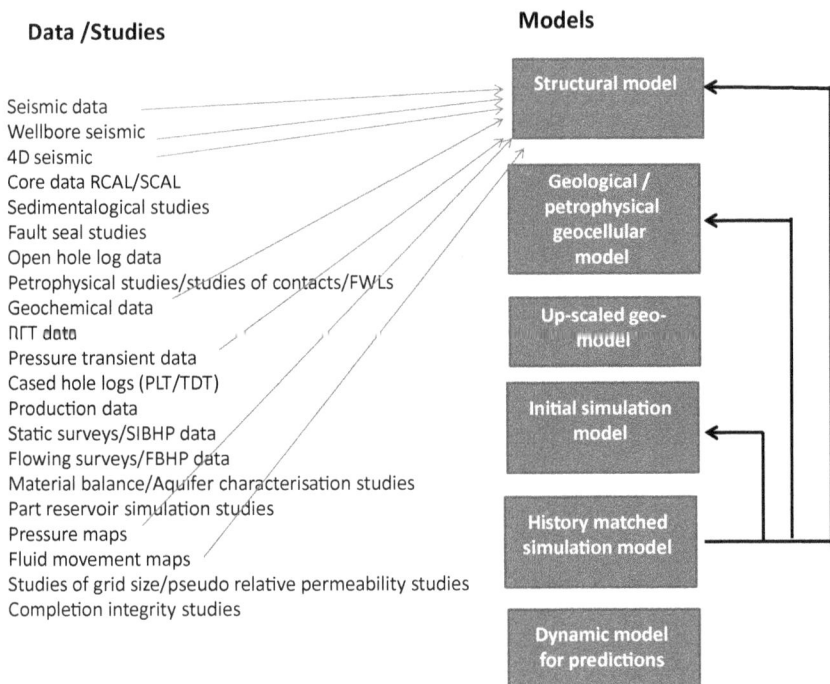

Figure 4.2.1. Overview of a typical reservoir modelling exercise.

Figure 4.2.1 shows the sequence of models that would normally be produced in the course of a reservoir modelling exercise. Also shown is an (incomplete) list of data and supporting activities that would be used in the modelling work. An attempt has been made to relate these activities to one of the models, the structural model. As can be seen, a wide range of data and studies contributes to the structural model, including both "static" and "dynamic" data, as discussed in the next section. It may be instructive to attempt to do this for the other modelling stages.

Figure 4.2.1 also shows the styles of iteration that may be associated with the history matching process.

Firstly, the initial simulation model may be "edited". This is entirely appropriate for some types of changes that would be introduced during the history matching process; examples include making changes to estimated rock compressibility, making global changes to permeability, and changing fault transmissibilities. For other cases, for example changing the style of permeability variation, it may be an acceptable way of exploring how a better match could be achieved but would be unsatisfactory as a final model.

Secondly, we may wish to update the geocellular model in a more systematic way than simply editing the simulation model input parameters. This could involve generating a different realisation of the geological model, exploring a different permeability model, or introducing a greater degree of "determinism" into the geological model.

Thirdly, we may wish to rebuild the structural model. This may be done in order to better model connectivity in the reservoir. There is currently less tendency to change the structure than other elements of the geological model during history matching. This is not a reflection of structural uncertainties being less important than, for example, uncertainties related to the sediment logical model; it is a reflection of the time-consuming nature of the structural modelling.

In all of the modelling stages illustrated in Fig. 4.2.1, consideration needs to be given to issues related to gridding. These are discussed later.

4.3. Static and Dynamic Data/Static and Dynamic Models

4.3.1. *Static and Dynamic Data*

There is a tendency to classify data as being "static" or "dynamic" and also to relate these data to the static (structural/geological) and dynamic (reservoir simulation) models that we build. It is not always straightforward to classify data in this manner.

Examples of "static data" include core descriptions and poroperm measurements on core data. Examples of dynamic data include production data and measurements of flowing and shut-in pressures in wells. Some data, for instance wire-line log data that can indicate water movement, would not fit easily into either category.

4.3.2.   *Dynamic Data and the Static Model*

For a producing reservoir there may be a lot of dynamic data available when the static geological model is being prepared. The extent to which these data are taken into account when constructing the static model will have a clear impact on subsequent history matching. It is generally considered good practice to account for some of these data in the static model. This is done with the hope that the process of history matching will be easier and that the resulting history matched model will be more useful.

An example is, using pressure transient test data to help in producing geologically based models for model permeability. Using this approach would make it likely that well productivity would be relatively well-matched in the simulation model prior to making any adjustments to the permeability. Changes made during the process of history matching may then be relatively modest and should not detract from the geological plausibility of the permeability model.

An alternative would be to build the initial model for permeability based solely on estimates from core. Such a model may require far greater modification to match historical data.

The following are examples of how dynamic data could be used in construction of the static model:

- use of pressure transient data, production log data and well productivity data to constrain the permeability model;
- use of RFT data and data on fluid movement from well logs to provide relatively deterministic models of the extent of baffles or barriers;
- use of pressure transient data/interference test data to help refine the structural model — fault locations and initial estimates of transmissibility of faults;
- use of geochemical data or pressure maps to better understand reservoir compartmentalisation;
- use of maps/cross-sections/fence diagrams illustrating fluid movement (produced by production geology studies) to help better define the geological model.

The use of these data holds out the prospect of the initial geological models being "closer to reality" than would otherwise be the case. This work is potentially very time-consuming but there is the hope that the resulting history-matched models will be more useful.

4.4. Issues Related to Reservoir Simulation/Up-Scaling

When comparing simulator results with observations it is important to have an understanding of some of the approximations/limitations that may be inherent in the use of a simulation model. The comments in this section assume that a conventional finite difference simulator is being used.

4.4.1. *Issues Related to Grid Size/ Numerical Resolution*

Issues related to grid size, numerical resolution and up-scaling will be relevant to a history match exercise. There will be an unavoidable lack of accuracy in the way that flow is represented with a relatively coarse grid model. Single- or multi-phase up-scaling is not able to fully capture all the details of flow at finer scale lengths than are captured in the coarse model. It may be better to estimate and accept this level of inaccuracy rather than to try to match historical data by changing physical inputs (for instance relative permeability) to compensate for numerical artifacts.

4.4.2. *Issues Related to Representation Rate Variation With Time*

Frequently (for reasons of computational speed or because of how production data are stored) simulation studies assume that well production and injection target rates can be set to (for example) monthly average values. It may also be convenient (computationally) to "write" simulator output at regular intervals and to use this for comparing historical and predicted values. This is relevant to pressure and saturation data that we may wish to compare with values from, for example, RFT and Thermal Decal Time (TDT) logs.

What effect does this choice of time intervals have on the comparison of observed and simulated data? We may expect that saturations would not change rapidly over short periods. Comparison of observed and simulated saturations would not be influenced too much by this choice of time intervals.

This may not be the case for pressures. Pressure communication may be relatively rapid over inter-well distances. The time variation (over periods of less than a month) can have a significant impact on pressures at an off-set well. This may impact how RFT data are modelled. We may want to represent the time variation of well rates in more detail prior to the time when RFT data is available. For similar reasons there would be a preference for comparing RFT data with pressures at the exact time the data were acquired rather than from simulator data stored at (for instance) monthly intervals.

4.4.3. *Issues Related to Representation of Well In-Flow in the Simulator*

Reservoir simulators do not generally represent the convergent flow to wellbores in detail. An approach developed by Peaceman 1978 is used in most commercial simulators to relate grid block and flowing bottom hole pressures. This approach was developed assuming 2D flow, a regular grid and uniform reservoir properties but is used fairly generally. As long as the simulated well rates are close to the actual rates this should allow flowing BHP data to be directly compared to simulated data.

Peaceman (1978) also described how, under the same conditions, grid block pressure could be related to pressures recorded during a pressure buildup. This approach can be used to compare grid block pressures with pressure data acquired during relatively short buildups. The shut-in and buildups will not be simulated directly. Static pressures acquired when wells have been shut in for long periods of time may be compared directly to the calculated wellbore (or indeed grid block) pressures.

A frequently adopted (if somewhat lazy) approach is to plot available shut-in pressure against calculated bottom hole and grid block pressures. Adding averages of grid block pressure from cells adjacent to the well and production rates can make such plots very useful.

4.5. Details of a "Conventional" Deterministic History Match Study

4.5.1. *Preparatory Work — Definition of Aims*

Clearly the first element of the study should be a definition of the (business) aims of the study. These aims then need to be translated into plans (budget, resources, timeline, etc.) for a technical study. It is important that those involved in the study remain aware of the business aims and are prepared to recommend stopping the study if those aims prove to be unrealistic.

4.5.2. *Data Review/Well Histories*

The initial technical element of the study should involve a review of the available data. This should be aimed at addressing two issues. First is the quality of the available data for reservoir characterisation; in particular are there any "gaps" that data acquisition could help fill. Second is the need to QC the data (especially the production and surveillance data). This will allow us to exclude bad data and to form a view as to the level of accuracy of the remaining data.

Consider shut-in Bottom Hole Pressure (BHP) data as an example. A review of the data may suggest that some data should be rejected because of:

- large difference between pressures from different gauges;
- pressure gradient in the wellbore not consistent with fluids in the wellbore;
- indication of leak in the well;

• flowing pressure reported as a shut-in pressure.

The data review would also give an estimate of the accuracy of the observed data that is being compared to the simulation data. This would involve:

• review of gauge accuracy;
• review of the accuracy of correcting data from gauge depth to the depth of the perforations;
• review of the accuracy of well deviation and wire-line depth measurements.

This process can be quite time consuming.

One area where particular attention should be paid is the analysis of cased hole logs. These can provide particularly valuable information for a history match (estimates of flow profiles from production logs, estimates of *in situ* saturation from pulsed neutron logs (PNL) and through casing saturation logs). The utility of these logs is, however, dependent on their giving estimates that are representative of conditions in the reservoir (Fig. 4.5.1).

Here a previously "dry" low pressure interval has a PNL response that suggests water invasion as a result of cross-flow. Clearly, being able to match saturations from PNL data should be very useful in a history match. There is, however, a need to ensure that these data are representative of conditions in the reservoir and conditions that the simulation model would reasonably be expected to replicate (it would be unusual to attempt to simulate explicitly a well being shut in for a short period before running a PNL).

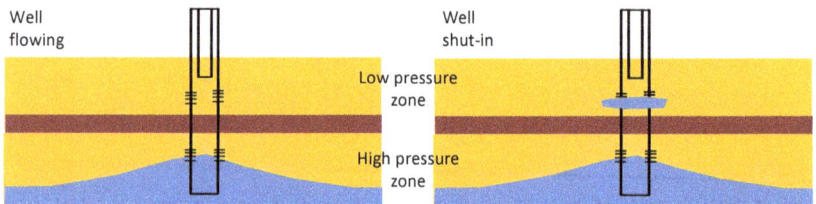

Figure 4.5.1. Illustration of a common problem with PNL interpretation.

A major element of any data is the preparation of well histories. Well histories would be expected to include the following:

- a review of drilling/logging the well;
- perforation and completion history of the well complete with all completion schematics;
- a review of cement quality;
- review of all work overs;
- comments on any mechanical problems encountered during the history of the well;
- reviews/plots of surveillance and production/injection data.

The data required for a history match would place more emphasis on historical data (e.g. old completions, history of mechanical problems) than would be needed for continuing operation.

4.5.3. *Preparatory Work — Data Acquisition Opportunities*

As we have seen, lack of uniqueness may be a major concern in a history matching exercise. Following a review of the available data it is appropriate to ask if additional data or processing would add value to the exercise. Clearly there may be significant time constraints in this process.

Firstly, we may require additional data that would reduce the uncertainty in model inputs. An example of this could be performing more core flood work to reduce uncertainty in residual oil saturations. Secondly, we may try to acquire data with the specific intention of attempting to match the data in the history matching exercise. An example of this could be acquiring enough pressure data to map pressure variation over the field.

4.5.4. *Preparatory Work —"Classical" Reservoir Engineering Calculations/Part Field Models*

Classical reservoir engineering calculations and "part field" simulation models are a useful precursor to a history matching exercise. They can help gain insight into which parameters are likely to have a significant influence on reservoir performance. They can also give insight into what limitations a full field simulation model may have due to numerical resolution.

Material balance calculations may usefully be carried out prior to history matching. This may allow a preliminary characterisation of aquifer properties and an assessment of whether the observed average pressure data are consistent with the assumed reservoir volumes and compressibility values.

4.5.5. *Review of the Simulation Model (and Geological Model)*

In the workflow that was described earlier, a review of the reservoir simulation model prior to commencing the history matching phase would be a part of many companies' QC procedures.

Many of the QC checks are relevant to history matching.

- QC of the basic engineering data to the simulation model will help for views on the accuracy of such data and the extent to which it would be acceptable to change these data.

  Example of relevance to history matching — Reviewing the pore space compressibility data would help define an acceptable range of values that could be used in the history match.
- QC of single- and multi-phase up-scaling should help define limits on how well the (relatively coarser) full field model will replicate the performance of finer simulation models. It could also involve an assessment of the limitations of the geo-model and the extent to which the up-scaled properties could be reasonably varied.

  Examples of relevance to history matching — The extent to which permeability values are uncertain could be assessed, as well as the extent to which numerical resolution could influence how fluid movement is modelled. We would want to avoid having unreasonable expectations as to how well we could match, for instance, the water cut of a well that had perforations only a few grid blocks away from a water zone in the simulation model.

- QC of how the structure is represented and how connectivity is represented in the simulation model can be important. If fault geometry/juxtaposition is important in determining flow paths in the reservoir then it is important to form a view of how uncertain this is and of the scope for connectivity being distorted in the up-scaling process.

  Example of relevance to history matching — Current workflow makes it relatively easier to produce many models for the variation of petrophysical properties (porosity and permeability) than the structural model. This puts a significant premium on having a "good" structural model and on ensuring that dynamic data have been considered when the structural model has been constructed.
- Review of how the wells are represented in the model.

The following may be of particular relevance to history matching.

- Are the wells correctly positioned with respect to faults in the model?
- Are the intervals that are open to flow correctly positioned with respect to baffles or barriers to flow?
- How well is the permeability height product seen in pressure transient analysis in agreement with the appropriate model values?
- Where well performance data indicate a mechanical "skin", is this represented in the initial simulation model?

This review can be relatively time-consuming.

4.5.6. *Outline of Approach to Matching the Model*

The major steps of matching a single model of a reservoir to historical data could be as follows:

(1) Preliminary simulation runs. The aim of these would be to ensure that historical well injection and production rates can be achieved. At this stage some effort may be needed to ensure that average reservoir pressures are very broadly matched.

(2) A review of:

  (a) the extent to which model input parameters could be reasonably changed;
  (b) expectations as to how much influence such changes would have on the simulation;
  (c) the extent to which model outputs should be matched.

(3) Based on the above, carry out a series of sensitivity studies to better understand the relation between model inputs and outputs.

(4) Attempt to match the model. This may go through a series of phases:

  (a) early efforts may be on broadly matching pressure;
  (b) next we may wish to broadly match fluid movement;
  (c) next we may wish to start matching pressure and fluid movement/production in more detail and to attempt to match well performance.

  The extent to which model inputs are changed in this process has to be constrained, either quantitatively or qualitatively, by considerations of "reasonableness". We also need to form, either quantitatively or qualitatively, a view on how well the data are matched. During this process, especially if some features of reservoir performance prove especially difficult to match, we may choose to review either the reliability of our measured data or the assumptions of geological modelling.

(5) Further refining of well performance (especially at the end of the history) and ensuring the model accounts "smoothly" for the transition between history and prediction.

(6) At this point it would be good practice to have a QC exercise to assess the fitness for purpose of the history-matched model.

4.5.7.   *How Well Should We Aim to Match Data?*

In this section, we will address the issue of how well we need to match data. The issues of what data we need to match will be addressed later. This will be done with reference to an example

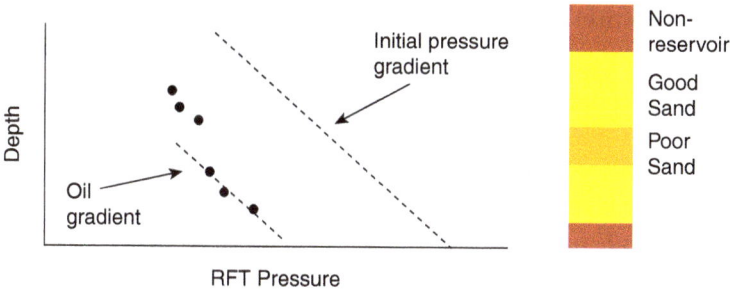

Figure 4.5.2. RFT data and reservoir rock quality.

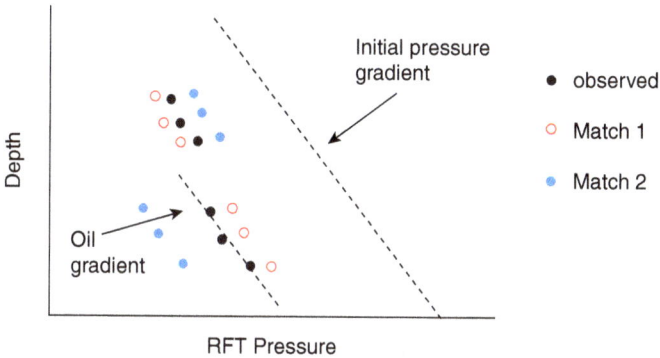

Figure 4.5.3. Possible matches to RFT data.

involving matching RFT data. Figure 4.5.2 shows some RFT data taken from the oil leg of a reservoir. Next to the RFT data there is an indication of reservoir quality.

There are a number of considerations we need to address when deciding what a good match to the data could be. These include:

- What is the measurement error of the RFT data? This would represent a lower limit of how well we may wish to match the data.
- To what extent would considerations of numerical resolution, choices of how to represent rate variation, etc. influence the quality of a match that could be expected? Again this would represent a lower limit of how well we would wish to match that data.

- Which feature of the pressure variation do we wish to match? If our main interest is in matching the general decline in the reservoir pressure then we would wish to match pressures to some fraction of the difference between initial pressure and current pressure. If, however, we wish to try to better understand the properties (possibly the vertical permeability or the lateral extent) of the poor quality sand in the above figure, then we may be more interested in matching the pressure "break" over the poor quality sand.

The aims we have in matching the data could influence what would be considered a better fit to the data. Consider the two potential "matches" to the RFT data shown in Fig. 4.5.3.

If we were interested in getting a good match to the average pressure decline, then Match 1 would be preferred. If we wanted a match that honoured the pressure break, then Match 2 might be preferable.

4.5.8. *Assessing the "Goodness" of a Match*

In the above, approaches to making assessments of the "goodness of fit" were discussed. It is possible to make this assessment qualitatively — for instance on the basis of visual comparison of observed and simulated data. (In doing this we may wish to use error bars to visualise the size of uncertainties in the observed data.) This approach is feasible and, at least for reservoir with a relatively small number of wells, is widely used.

For reservoirs with a relatively large numbers of wells, or if we wish to use any computer-assisted approaches to history matching, it is necessary to quantify how well the data are matched. A common approach would be to use a weighted Root Mean Square (RMS) error:

$$\sum w_i(x_i(\text{measured}) - x_i(\text{simulation}))^2/\sigma_i^2, \qquad (4.5.1)$$

where the w_i are weights and the σ_i represents a target level for fitting the data. Clearly in using such an approach care needs to be taken in choosing appropriate w_i and σ_i values.

This only measures how well or badly observed data are matched. We may also want to account for how realistic or unrealistic the changed input parameters to the simulation model are. This is discussed by Schulze-Riegert and Ghedan (2007).

4.5.9.   *Well Controls During the History Match*

The available approaches to controlling production wells during the history period include:

- control on oil (or gas) rate;
- control on liquid rate;
- control on reservoir volume rate.

If a perfect match is achieved then all of these methods will give the same result because production of all the phases (and pressure) will be matched. The choice of control mode should be dictated by the way this influences our ability to match the model.

There are two arguments in favour of using reservoir volume rates. The first relates to matching the reservoir pressure. Consider the material balance equation for a reservoir (or for a fault block within a reservoir) and for convenience assume there is no free gas. The material balance equation is:

$$NpBo + WpBw = NBoi[(Bo - Boi)/Boi$$
$$+ (cwSwc + cf)/(1 + Swc)\Delta P] + WeBw.$$
$$(4.5.2)$$

Assuming constant oil compressibility this becomes,

$$NpBo + WpBw = NBoi[co + (cwSwc + cf)/$$
$$(1 - Swc)]\Delta P + WeBw. \qquad (4.5.3)$$

The left-hand side of this equation is the reservoir volume production. If we get this, the stock tank oil initially in place, fluid compressibilities, and the aquifer influx term (*We Bw*) correct, then we should get the correct pressure drop. This would suggest that it should be

easier to match pressure by carrying out a series of simulation runs
where the reservoir volume production is "correct".

The use of reservoir volume production targets may also make it
easier to match water movement. Consider the case of trying to match
simple 1D flow by adjusting reliable permeability curves. Figure 4.5.4
shows water cut development for three cases:

(1) a "correct case";
(2) a "modified case" with a slightly different reliable permeability
    where reservoir volume production is the same as in the "correct"
    case;
(3) as above, but where the oil production rate is the same as in the
    "correct" case.

In the modified case water breakthrough is later. For the case where
oil rate control is being used, the reservoir volume rate decreases
causing a further delay in breakthrough time. The difference becomes
greater at later time due to the water cut being underestimated and
the reservoir volume rate too low.

Similar considerations would lead to the use of oil rate control,
exaggerating the difference between the "true" and "modified" cases
if the modified case had earlier water breakthrough.

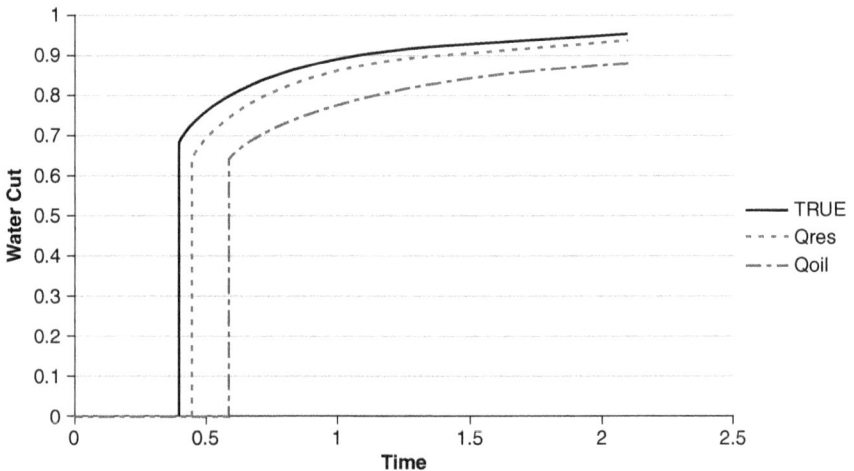

Figure 4.5.4. Water cut development with time.

Based on the above, the use of reservoir volume rates to control production would seem to be highly desirable. The benefits would be:

(1) to make it easier to match pressure, and also possible to match pressure without matching the production of individual phases;
(2) to make the process of matching water (or gas) production more "stable".

We also need to make decisions about how to control injection wells. For water injection, due to the small compressibility, it should not make much difference whether the injection is controlled by reservoir or surface rate.

What about gas injection? If reservoir volume rates were to be calculated based on a pressure that was too low, then the rate, in terms of surface volumes, would be reduced. The converse would apply if the pressure were too high. This could suggest that it would be preferable to use surface volume constraints.

4.5.10. *Initial Simulation Runs*

Initial runs of the simulation model and some work to ensure the simulation model correctly accounts for perforation histories may be appropriate prior to any sensitivity studies. This work should have the following aims:

(1) Review average reservoir pressures. If there is a very large mismatch between simulated and actual reservoir pressure it is desirable to attempt to better match history (by adjusting aquifer influx of fault transmissibility for instance).
(2) Ensure that wells are able to produce at their historic rates.
(3) Review production data to see if this indicates any errors in the perforation history.

If it is not possible to get a very approximate match to reservoir pressure then the model may not be an appropriate "base case" for sensitivity studies. Also, if pressures are too low then it may prejudice our ability to produce wells at their historical rates; the PI of the completion multiplied by the maximum possible pressure drop may not be high enough.

If the model has a broadly correct pressure but some wells are unable to produce (or inject) at historical rates it would suggest either of the following:

- the kh products are significantly in error;
- the perforated intervals are incorrect;
- there is flow "behind pipe" that is not captured in the model;
- a well has been stimulated (negative mechanical skin?) and this is not reflected in the model.

4.5.11.   *Review of Scope for Changing Model Input Parameters*

Before proceeding with a history match it is desirable to review (and document) the input parameters that could be varied and the extent to which it would be acceptable to change them. Clearly the types of changes will be very case specific. It may be appropriate to give some examples.

Example 1 — Rock compressibility

Based on a review of experimental (SCAL) data and of uncertainties about stress regimes we could form a view on how uncertain rock compressibility data are. This could give us:

- a most likely value (or values);
- a range of values that would be acceptable;
- a larger range that would be considered possible but less likely.

Example 2 — General level of uncertainty in permeability

Based on reviews of the geological model, up-scaling methods, and pressure transient data we should be in a position to form a view about the accuracy of the estimated average level of permeability in the model. This could give us a range of "multiplication factors" that could be applied, either globally or locally.

Example 3 — Local barriers to vertical flow

Based on geological studies we may recognise that there could be local barriers to flow. These could, for instance, be shales or cemented layers. The existence of these at wells may be recognised from well logs and they may be accounted for to some extent in the geological model (and through that in the simulation model). We may wish to form a view on whether introducing (or removing) barriers or baffles to vertical flow in the model should form a part of the history matching process.

4.5.12. *Sensitivity Study*

Prior to carrying out a "manual" history match (also before or as part of a "computer-assisted" match) we may want to carry out a sensitivity study. In this study we would look at "inputs" that were uncertain and evaluate the effect of making "reasonable" changes to them on the simulation results. Choices could be along the lines of those in Table 4.5.1; it should be stressed that this is just an example.

Table 4.5.1. Illustrative inputs on uncertainty/sensitivity.

Input	Sensitivity	Discussion
Permeability	± 50%	Based on comparisons of model permeability values to estimates from PBU (Pressure Build Up)
kv/kh	× 10 and × 0.1 of base case	Based on kv/kh estimates from PBUs and fine scale models
Krw	× 0.5 and × 1.5	Based on analysis of SCAL data
Bo (the oil formation volume factor) and oil viscosity	No sensitivity	PVT (Pressure, Volume and Temperature) data give us good estimates
Fault seal	Complete seal to no seal	We may have very little understanding of fault seal potential
Etc.	Etc.	Etc.

If models have been produced that have a significant stochastic input it may be valuable to compare results from different model realisations at this point.

4.5.13. *The History Matching Process*

In a "manual" history matching it would be common to adopt the following approach.

- Concentrate first on pressure, then on fluid movement, then on details of well performance.
- Concentrate first on matching at a field level (e.g. field average pressure, field water cut), then in more detail.
- First change inputs globally; later make more local variations.

Clearly the approach to matching the data is iterative. For example

- if we change relative permeability to help match water movement then this will change fluid mobility and hence the pressure match;
- changing the "skin" term on a well to better match well flowing BHP could change the proportion of in-flow from different layers; this could in turn change both the fluid movement and the pressures.

However, this approach is based on the following observations:

- The time scale for pressure communication over reservoir scale lengths will be much shorter than the time scale for fluid movement. Thus matching pressure data is likely to be more useful in constraining large scale aspects of the reservoir model.
- Pressure gradients will determine the general direction of flow. There would be little point in trying to match the detail of fluid movement before the pressure was broadly matched.
- Many of the inputs that could be changed to better match fluid movement (reliable permeabilities, local changes to permeability, modest changes to kv/kh, etc.) may have little effect on the match to the pressure.

4.5.14. *Matching Pressure*

The data that we may wish to match include:

- shut-in bottom hole pressure (SIBHP) data,
- flowing bottom hole pressure (BHP) data,
- RFT data,
- interpreted reservoir (or fault block) pressures,
- pressure maps.

RFT data may be especially useful because of their relatively high accuracy and their ability to give information of baffles/barriers to vertical flow.

The inputs we would expect to have a major impact on pressure would include:

(1) aquifer properties,
(2) volumes in place, fluid and pore space compressibility,
(3) bubble point pressure/variation of bubble point spatially (for reservoirs where pressure falls below the bubble point),
(4) general level of permeability within flow units,
(5) fault transmissibility,
(6) kv/kh and the transmissibility/lateral extend of barriers or baffles to lateral flow,
(7) the split in production and injection rates between different flow zones.

In many cases, it may be appropriate to match estimates of average reservoir pressure early in the matching process. This could involve mainly (1)–(3) above points of.

Matching pressure gradients can be very valuable. If we are to correctly capture the direction of flow in the reservoir then we need to match pressure gradients. This may be influenced by point (1) (especially how any aquifer is connected to the model) and points (4) and (5) of above.

If we are interested in matching pressure variation between flow units and in better understanding vertical communication, then we would also be interested in points (6) and (7). For point (7), matching the split in flow between could involve attempting to match PLT (Production Logging Tool) data. Thus the process of matching pressure may involve attempting to match the performance of individual wells in some detail.

4.5.14.1. *Pressure match example*

Consider a very simple reservoir model as illustrated in Fig. 4.5.5 below:

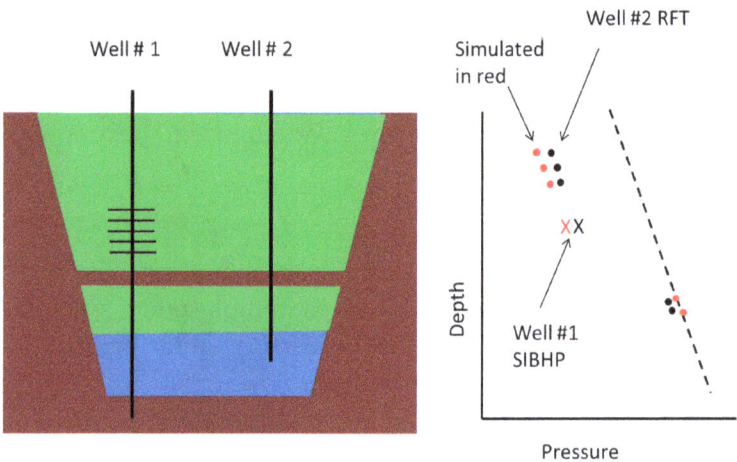

Figure 4.5.5. Conceptual reservoir cross-section and RFT data.

In this case there is an isolated reservoir (surrounded by shale — shown in brown) with oil and water columns and a shale, believed to be a barrier to vertical flow, within the oil column. The first well is drilled and is produced from above the shale barrier so as to avoid water production. A second well is drilled and RFT data acquired. At the same time as this a SIBHP survey is carried out for the first well.

Results are compared with a simulation model. This comparison shows the following:

- simulated pressures are a little too high above the shale barrier;
- there is a decline of pressure below the shale barrier that cannot be easily explained.

How could we approach history matching our model? We can look at two aspects of this. Firstly how can we explain the data from below the shale?

- How accurate are the estimates of initial pressures? And how accurate is the well #2 RFT data? Can we explain the data by measurement error/inaccuracy?
- Could there be some flow through the shale? We would need to review geological views on, for example, small faults allowing communication across the shale.
- Is there scope for flow from below the shale in well #1? This could be a result of poor cementing in the well. We may wish to review any cement bond logs. It may also be possible to run additional logs, for instance temperature logs, to try to see if there is flow of fluid "behind pipe".

Secondly how do we go about matching the pressure data?

If we assume that the pressure decline below the shale is real and not a result of flow behind pipe then matching the pressure data could involve changing inputs including the following:

- the volumes in place above and below the shale. This could involve changes in pore volume (porosity, net to gross or gross rock volume);
- fluid properties (principally oil and water compressibility values);
- rock (pore space) compressibility;
- the factors influencing flow across the shale (shale permeability, introducing one or more faults into the model to create sand-to-sand juxtaposition, etc.).

We would expect to be able to match the observed data in more than one way — the problem of non-uniqueness. For instance, changes in volumes in place and in fluid and rock compressibility will have very similar effects on the average pressure.

In terms of primary production, this non-uniqueness may not be much of a problem — think about a simple material balance equation. If we intend to water-flood the reservoir then there is a clear difference between having larger oil in place and having larger fluid or rock compressibility. Clearly in this case there is a benefit in reducing the "level of non-uniqueness" by reducing the uncertainty associated with compressibility values by having more and/or better experimental data.

4.5.15. *Matching Fluid Movement*

The data we may wish to match include:

- the production of oil, gas and water;
- open and cased hole logs that indicate fluid saturations;
- estimates of fluid distributions from seismic data — in some cases 4D seismic may provide very useful data;
- interpretations of fluid movement based on production geology studies — for example fluid movement maps and cross-sections/fence diagrams.

(Clearly we need to be mindful of the differences between measured data and interpretations. Attempting to match interpreted data, such as water movement maps, represents a way of building on the understanding of reservoir mechanisms developed by production geology studies.)

For simplicity the following discussion and the examples deal with two-phase oil–water problems. When considering the inputs that could be changed to better match, we should consider the following features of oil water displacement:

- the broad pressure gradients within the reservoir — these will determine the direction of flow in the reservoir;
- the factors influencing local displacement — these would include relative permeability and capillary pressure models;
- the factors influencing volumetric sweep efficiency — these would include local permeability heterogeneity, the mobility ratio of the displacement the stability of the displacement, and the tendency water to cone or cusp.

Assuming that there exists a broadly adequate match to the pressure, the inputs that could be changed to attempt to match water movement would include:

- the degree of permeability heterogeneity;
- permeability anisotropy (kv/kh);
- the relative permeability model (Kr end points, saturation end points, shapes of curves);
- properties of faults and fractures (local flow barriers or enhancement to flow, connectivity, etc.).

Visualisation of the flow in the reservoir model, for example by viewing a series of saturation plots over time, can be a useful tool when determining which of these changes are more likely to be appropriate. In particular, it can act as a check on whether the "style" of water movement seen in the simulation model is consistent with our understanding from other sources.

Other examples of visualisation tools could be comparisons of saturation maps based on simulation and from surveillance data and "traffic light" plots indicating (as a function of time) wells where water production is correctly predicted, over-predicted and under-predicted.

As noted previously, in attempting to match water production history we need to give consideration to the numerical resolution of the simulation model. We can look at two examples.

Example 1 — There is early water breakthrough in the simulator because there are only two cells between a cell that is below the oil–water contact (OWC) and a cell that is attached to the well. If we try to fix the problem by, for instance, reducing the end point water reliable permeability (Krw), then we risk severely distorting how we model flow everywhere this change is implemented.

Example 2 — A well cones water "in reality" but this cannot be captured in the simulation model because the lateral resolution of simulation grid is not fine enough. It may be possible to make the simulation model cone water, for instance by reducing permeability in the vicinity well (thus decreasing pressure near the perforations and promoting the upward flow of water). This would, however, severely distort well performance.

In both of these examples, there would be scope for increasing grid resolution close to the wells. If, however, this is not possible, then it may be better to live with having poor matches to water production (provided that we understand the cause) rather than distorting model inputs in a way that could result in unrealistic predictions.

4.5.16. *Matching Water Movement — Example 1*

The first example of matching water movement is illustrated in Fig. 4.5.6.

This also shows a comparison of the simulated and actual water production at the single production wells. The actual water cut evolution is shown by the solid line and the simulated evolution is shown in the dashed curve.

How could the model be changed to better match the observed production data? Two aspects of the mismatch of simulated data to history are worthy of note. Firstly, although both simulated and actual water cut development would indicate some layered behaviour (the water cut increases in two distinct phases), the simulated water

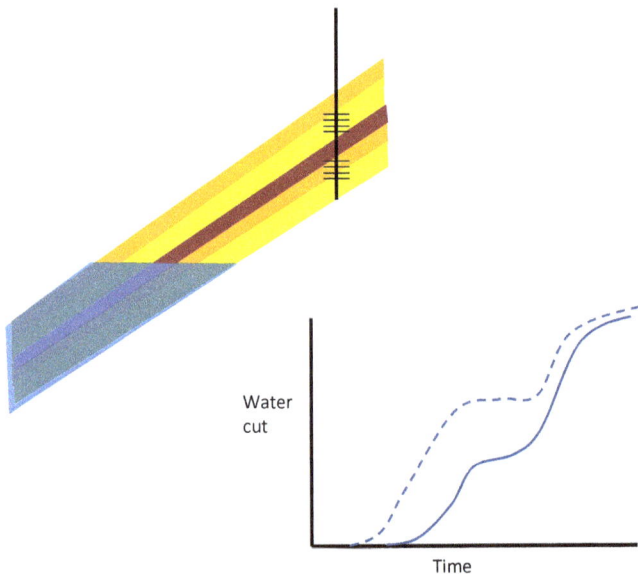

Figure 4.5.6. Conceptual reservoir cross-section + water cut development.

cut could indicate higher flow in the first layer to produce water. Secondly the simulation model produces water earlier than is actually the case.

When we seek to match the water cut we need to think about the other data that are available, e.g.

- PLT data that could give information of flow in different layers;
- PNL data that could give information of swept zone saturations.

What changes could be made to better match the water cut development?

The split kh product and φh product will influence the style of water cut development, as will the mobility ratio of the displacement. The timing of water production will be influenced by volumetric and local sweep efficiencies.

One approach to modelling that data could be as follows.

- If PLT data are available then attempt to match the split between layers by changing modifying layer permeability values (but keeping total kh in line with estimates from PBU data).
- Investigate how reasonable changes to reliable permeability end points and the shape of reliable permeability curves could influence the time of water production. The following could give later water production:

  o decreasing *Sorw* (and hence decreasing local sweep efficiency);
  o decreasing *Krw* — this would make the displacement less stable (it would tend to reduce volumetric sweep and local sweep efficiencies).

- Investigate how reasonable changes to permeability and porosity could change water production. The following could give later water production:

  o reducing permeability heterogeneity within the main flow units;
  o possibly increasing kv/kh;
  o increasing porosity.

Based on this work we can try to produce a reasonable match to the data. There are two potential problems.

Firstly, we may find that it is possible to achieve reasonable matches in several ways — the problem of non-uniqueness. For example, decreasing *Sorw* and decreasing *Krw* could both improve the match. One would have the effect of reducing the mobile oil volume and the other would not. These changes could lead to significantly different views on reserves.

Secondly, we may find that no set of "reasonable" changes to model inputs allows us to match the observed data. In this case we may need to review our assumptions. The following are examples of what we may do.

- Review the production data. Are we sure of the reliability of the water cut data for instance?
- Review the mechanical condition of the completion. Is there scope for water being produced from other intervals (e.g. an underlying aquifer) because of mechanical problems?
- Review our views on how far input parameters could be varied. For example if we found it impossible to match the production data without increasing local sweep would we want to review the SCAL data and the acceptable range of Sorw values?
- Review our basic geological input. For our present case we could envisage a barrier to lateral flow down dip of the production well delaying water production. It would be simple to test whether this would work in a simulation model. We would then have to review geophysical data to see if this is plausible and to see if it could be consistent with PBU data.

4.5.17. *Matching Water Movement — Example 2*

This is a variant of Example 1. The aim of this example is to illustrate the importance of matching pressure and hence the general direction of flow on matching water movement.

In this case there is the possibility, if the fault is not sealing, of oil influx from the fault block to the right of the figure (Fig. 4.5.7). Having flow across the fault in the simulation model would tend to delay water production in the well.

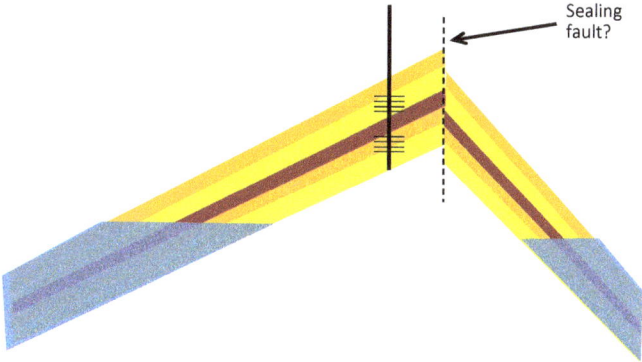

Figure 4.5.7. Conceptual reservoir cross-section.

If we have an incorrect understanding of pressure communication and the broad pattern of fluid movement, then there is scope for making inappropriate changes to match fluid movement. Clearly if we have more information, for instance PBU data giving information on the nature of the fault, then we would be better able to produce an accurate model.

4.5.18. *Matching Water Movement — Example 3*

In the previous two sections we have looked at changes that could be made to match water movement for a model where only a single well is being considered. What would we do in the case of attempting to match several wells?

Two broad approaches could be possible. Firstly we could attempt to achieve the best overall match to history by making global (or "regional") changes to selected parameters, as discussed above. Secondly, we may attempt to make many local changes to better match individual well performance. Both approaches have some virtues. The first is simpler and may, arguably, lead to a model with more predictive capability. The second will provide a better ability to model short term production prediction.

If the former approach is preferred then it may still be appropriate to investigate what sort of local model changes would enable performance to be better matched to individual wells and to ask whether these changes are in line without understanding the level of uncertainty in reservoir description.

If a decision is made to introduce local changes to match individual well performance then some changes may be more acceptable than others. Introducing a barrier to flow that is not seen on seismic data but which is indicated by PBU analysis may be acceptable. Introducing local changes to reliable permeabilities or local "pore volume multipliers" may be less acceptable.

4.5.10. *Matching Well Pressures/Detailed Well Performance*

Well in-flow or injection performance needs to be modelled. Also we may wish to match the details of in-flow or injection profiles, as seen for example on production logs.

An important element of this work would involve matching flowing BHP data. There may be particular emphasis on matching the most recent data. Calibrating the model to these data is important for using the model for prediction. We would also want to achieve a reasonable match to these data throughout the history period. If a well is producing from more than one flow unit with different pressures then changes to well productivity will change the split of off-take between the different units. This means that matching well performance cannot be decoupled from matching pressure and fluid movement.

The most straightforward changes would involve introducing "productivity index multipliers" for wells. Ideally these multipliers would be consistent with our understanding of how well the model matches kh products and our understanding of skin factors.

In some cases, mainly when the perforated interval is represented by a small number of cells, well productivity may be poorly represented in the model when there is multi-phase flow. There could be segregated flow in reality but diffuse flow, and reduced total mobility, in the model. In such cases exploring the use of "well pseudos (possible straight line curves) may be appropriate.

We would also wish to model well and tubing performance prior to using the model for prediction. If the model is doing a good job of matching flowing BHP and water cut then this matching should involve making at most rather minor changes to tubing performance curves.

4.5.20. *The Transition to Prediction*

It is generally considered appropriate to attempt to replicate the latter part of the production history using the well controls that will be used in a prediction case. This would involve running a simulation model with the wells controlled by Tubing Head Pressure (THP) and applying appropriate facility limits. If the simulation does a poor job then it would not be expected to be useful for prediction. If it does a good job it may be useful for prediction.

4.6. Automatic and Computer-Assisted History Matching/Multiple Matches

For many years there has been research into methods for automatic and computer-assisted history matching. A good review and set of references is provided in the literature of Schulze-Riegert and Ghedan (2007). Recently there have been significant advances and computer-assisted matches with commercially available software increasingly becoming relatively common.

The availability of automatic or computer-assisted matches, together with advances in geological modelling, have made producing multiple matched models a more realistic possibility. A degree of automation in the matching process has the benefit of making the process less subjective. It also may reduce the risk of a "manual" history matching process resulting in multiple models that were too similar.

4.7. Conclusions

Producing history-matched models may allow us to address important questions on reservoir development and management. The quality and utility of the models depends on how well dynamic data are accounted for in the modelling process. Much of the effort associated with accounting for the dynamic data should form a part of the construction of the geological model. This may make the process of history matching the dynamic model easier and the resulting model more useful.

The process of history matching should be performed with a view to making the resulting model fit for purpose. Giving emphasis to making only reasonable changes to model inputs rather than "forcing" outputs to fit observed data may help achieve this aim.

References

Mattax, C. C. and Dalton, R. L. (eds.) (1990). *Reservoir Simulation, SPE Monograph Series, Volume 13*, Richardson, Texas.

Peaceman, D. W. (1978). Interpretation of well block pressure in numerical reservoir simulation, *SPE Journal*, **18**(3), 183–194.

Schulze-Riegert, R. and Ghedan, S. (2007). Modern Techniques for History Matching, *Proceedings of the 9th International Forum on Reservoir Simulation*, 9–13 December, Abu Dhabi, UAE.

Index

www.ingramcontent.com/pod-product-compliance
Lightning Source LLC
Chambersburg PA
CBHW050551190326
41458CB00007B/1996